Palgrave Studies in the History of Science and Technology

Series Editors

Designed to bridge the gap between the history of science and the history of technology, this series publishes the best new work by promising and accomplished authors in both areas. In particular, it offers historical perspectives on issues of current and ongoing concern, provides international and global perspectives on scientific issues, and encourages productive communication between historians and practicing scientists.

More information about this series at
http://www.springer.com/series/14581

William deJong-Lambert • Nikolai Krementsov
Editors

The Lysenko Controversy as a Global Phenomenon, Volume 1

Genetics and Agriculture in the Soviet Union and Beyond

palgrave
macmillan

Editors
William deJong-Lambert
Bronx Community College
City University of New York
New York, USA

Nikolai Krementsov
University of Toronto
Ontario, Canada

Palgrave Studies in the History of Science and Technology
ISBN 978-3-319-39175-5 ISBN 978-3-319-39176-2 (eBook)
DOI 10.1007/978-3-319-39176-2

Library of Congress Control Number: 2016962321

Printed on acid-free paper

This Palgrave Macmillan imprint is published by Springer Nature
The registered company is Springer International Publishing AG
The registered company address is: Gewerbestrasse 11, 6330 Cham, Switzerland

Preface, vol. 1

In volume one of *Lysenkoism as a Global Phenomenon* we focus on events in the Soviet Union, chronicling Trofim Lysenko's precursors, rise to prominence, and scientific and cultural influence. The general introduction outlines the history, historiography, and future direction for studies on the Lysenko controversy as a global phenomenon. The first chapter, "Lysenko's Predecessors: the Demichinskiis and a New Technique of Cereals Cultivation," describes Nikolai Alexandrovich and Boris Nikolaevich Demchinsii's development of a new technique of cereal production, similar to Lysenko's later innovation, vernalization. The author asks why the Demichinskii's "panacea" for Russia's perennial food shortages was rejected while Lysenko's was adopted, to show the role of the scientific community in promoting Lysenko's career.

This line of inquiry extends into the next chapter, "State Officials and Would-Be Scientists: How the Ukrainian Ministry of Agriculture Discovered for Lysenko that He Had Made a Scientific Discovery," where the author challenges the conventional wisdom that Soviet press played an important role in Lysenko's celebrity and success. Rather, the press was simply reiterating the views of officials at the Ukrainian Ministry of Agriculture, who had their own motives for favoring Lysenko.

The next chapter, "Pavel Pantelimonovich Luk'ianenko and the Origins of the Soviet Green Revolution," describes Luk'ianenko's success in developing high-yielding semi-dwarf wheat. That Luk'ianenko was able to conduct his research, even while Lysenko was in power, is evidence that Lysenko's anti-genetics campaign was not quite as extensive or comprehensive as previously understood.

The last chapter in this volume, "Lysenko's 'Michurinism' and Art at the Moscow Darwin Museum, 1930s–1950s," chronicles how Lysenko's influence was reflected in the exhibits of the Natural History Museum of Moscow. In this reading, Lysenko's Michurinism provided a resource for survival and access to state resources. Read along with the chapters described above, we gain a more nuanced understanding of Lysenko's impact and career in his homeland. The next volume describes how this influence spread and manifested in variety of nations on both sides of Iron Curtain while he was in power.

Contents

Notes on Contributors

Luis Campos is Associate Professor of History of Science and Senior Fellow at the Robert Wood Johnson Foundation Center for Health Policy, University of New Mexico. His scholarship integrates archival discoveries with contemporary field-work at the intersection of genetics and society. He is the author of *Radium and the Secret of Life* (University of Chicago Press, 2015), a prehistory of radiation genetics. He is currently at work on his next book on the history of synthetic biology.

Francesco Cassata is Assistant Professor of Contemporary History at the University of Genoa. He has published several books on the history of eugenics and scientific racism in Italy, on the history of Lysenkoism in Italy, on the Italian geneticist Adriano Buzzati-Traverso and the International Laboratory of Genetics and Biophysics in Naples (1962–1969). He is currently working on several research projects: the history of EMBO-EMBC-EMBL; the anthropological studies of Dante's bones from 1865 to 1938; the radiogenetic and mutagenesis program of durum wheat at the Nuclear Energy Research Centre Casaccia (Rome), from the 1950s to 1980s.

Olga Elina is a Leading Research Fellow at the Institute for the History of Science and Technology of the Russian Academy of Sciences in Moscow. She is the author of a two-volume history of Russian agricultural science and has published numerous articles on the subject in leading international journals.

William deJong-Lambert is an Associate Professor of History at Bronx Community College, CUNY, and Associate Faculty of the Center for Science and Society, Columbia University. His publications include *The Cold War Politics of Genetic Research: An Introduction to the Lysenko Affair* (Springer 2012), a co-edited special issue on Lysenkoism of the *Journal of the History of Biology* (45/3, 2012), as well as several articles in *Journal of Cold War Studies*, *Historical Studies in the Natural Sciences*, *Studies in the History of Biology*, *East European Politics and*

Societies and others, as well as chapters in edited collections published by Amsterdam University Press, Routledge and Neriton. His research has been funded by the National Science Foundation, Fulbright, the American Council for Learned Societies, the American Philosophical Society and other organizations.

Lukas Joos has completed his MA at the University of Zurich in 2012. He is currently working on preparing for publication his thesis on "The rise of T. D. Lysenko to the presidency of VASKhNIL in light of its coverage in the newspaper *Pravda*."

Eduard I. Kolchinsky is the director of the S.I. Vavilov Institute for the History of Science and Technology, St. Petersburg, Russia. He has published and edited numerous books on the history of Russian biology, philosophy, and evolutionary thought.

Nikolai Krementsov is a Professor at the Institute for the History and Philosophy of Science and Technology of the University of Toronto, Canada. He is the author of several books and numerous articles on the history of Soviet science, biomedical research, eugenics, and intersections of science and literature.

John Marks is Associate Professor in French Studies in the Department of French and Francophone Studies at the University of Nottingham, England. His main research area is contemporary thought and social theory. He is currently working on a study of discourses of life in postwar French thought on biology.

Miklós Müller Upon his retirement in 2005 as the head of the Laboratory of Biochemical Parasitology at the Rockefeller University, Prof. Müller turned his interests to the history of twentieth-century biology. His studies concern two topics: the life and works of Hungarian theoretical biologist Ervin Bauer (1890–1938) who successively worked in Hungary, Czechoslovakia, Germany, and the Soviet Union, and the impact of Soviet distortions of biology (Lysenko and Lepeshinskaya) on biology in Eastern Europe in the 1940s and 1950s.

Cristiana Oghina-Pavie is an Assistant Professor in the Department of History of the University of Angers, France, and member of Centre de Recherches Historiques de l'Ouest CERHIO UMR CNRS 6258. Her research interests concern two main themes: history of international relations and life sciences. She studied French-Romanian acculturation of models of nation, state, and politics. She currently works on relationships between horticulture and plant sciences in the ninteenth and twentieth centuries with focus on physiology and heredity. With epistemologists and biologists, she develops interdisciplinary approaches to the effects of breeding on the genetic diversity of cultivated roses.

Gábor Pallo is Senior Consultant at the Visual Learning Laboratory, Budapest University of Technology and Economics. His fields of research include history of chemistry and physics, twentieth-century history of natural sciences in Hungary, philosophy of science, history of migration of scientists, the relationship between science, politics, and philosophy. Some recent publications: *Zsenialitás és korszellem* (Genius and Zeitgeist), (Budapest: Áron Kiadó, 2004); "The Advantage and Disadvantage of Peripheral Ignorance: The Gas Adsorption Controversy," *Ambix*, 57, 2, (2010): 216–30; "Early impact of quantum physics on chemistry: George Hevesy's work on rare earth elements and Michael Polanyi's absorption theory," *Foundations of Chemistry*, 13, 1, (2011): 51–61.

Hirofumi Saito has a PhD (2008) from the Tokyo Institute of Technology (Tokyo Tech). From 2009, he has been working as a research associate at the Tokyo Tech, focusing on the history of Soviet genetics. Since 2010, he has been investigating documents of the Archive of the Russian Academy of Sciences. Recently, he published two works, which reconsider the process of Lysenko's intervention into the Institute of Genetics in the late 1930s and Lysenko's personal state in his agrobiology school during World War II.

Pat Simpson is Reader in Social History of Art and Research Tutor, at the School of Creative Arts, University of Hertfordshire, UK. She has published extensively on Soviet art and culture in *Russian Review*, *Art History*, *Oxford Art Journal*, *Sculpture Journal*, and the *International Journal of the History of Sport*. She has also contributed chapters to the anthologies *Art, Sex and Eugenics* (2008), *The Art of Evolution* (2009) and *Regeneration and Degeneration* (2015). Her chapter is based largely on research funded by a British Academy grant 2010–2011. Pat's forthcoming book, *Art and Evolutionary Biopolitics*, is in preparation for 2017.

Mark B. Tauger is an Associate Professor of history at West Virginia University. After earlier work in musicology, he completed his PhD at UCLA and conducted research in Russia and Ukraine with support from IREX. He has published extensively on Soviet famines and agriculture, the Bengal famine in India in World War II, world agricultural history, and other areas. His work has won the Eric Wolf Prize of the *Journal of Peasant Studies* and the Rasmussen Prize of the Agricultural History Society.

List of Figures

"Lysenkoism" Redux: Introduction

Nikolai Krementsov and William deJong-Lambert

Soviet agronomist Trofim Denisovich Lysenko (1898–1976) is one of the most notorious figures in the history of twentieth-century biology. US biologist and writer Stephen Jay Gould has described Lysenko's August 1948 announcement that the Central Committee of the Soviet Communist Party had approved his concepts of heredity and evolution as the "most chilling passage in all the literature of Twentieth Century science."[1] In 2015, many Russian biologists and agricultural scientists will celebrate (and some, judging by certain recent publications,[2] mourn) a "Golden Jubilee": the semi-centennial of the final break of Lysenko's nearly thirty-year-long hold on the development of their fields. Like Darwin's and Mendel's, Lysenko's name has been converted into an "ism," a label denoting fierce debates that surrounded his science and his career for more than half a century. Over the years, this label has been applied to a variety of individuals, ideas, practices, and images, making "Lysenkoism" not merely a shorthand for Lysenko's biological theories,[3]

N. Krementsov (✉)
University of Toronto, Toronto, ON, Canada

W. deJong-Lambert
Bronx Community College, CUNY, Columbia University, New York, NY, USA

W. deJong-Lambert, N. Krementsov (eds.),
The Lysenko Controversy as a Global Phenomenon, Volume 1,
DOI 10.1007/978-3-319-39176-2_1

but a textbook example of pseudoscience, the infringement of political authorities on academic freedom, and the corrupting influence of ideology on scientific culture.

A bibliography of publications on Lysenko and Lysenkoism would include more than 1000 items, ranging from short notes in daily newspapers to voluminous scholarly tomes. Yet, despite the extensive notoriety, there is still no full-length biography of the man, just a few short entries in various encyclopedias.[4] We know surprisingly little about his parents and siblings, his childhood and youth, his personal life, marriage, and children, his personal friends and tastes (in music, literature, food, arts, etc.), or his personal beliefs and attitudes. But the major events of his public life and the story of his "rise and fall"—the *Lysenko Affair*, in David Joravsky's astute characterization—are well-documented.[5] In contrast, the historiography of writings on Lysenko and Lysenkoism—an analysis of the different ways in which his life and works have been presented and assessed by numerous commentators and scholars—remains largely an uncharted territory. The first part of this chapter will provide a sketch of Lysenko's life story. The second will analyze the major trends in writing on and approaches to that story developed over the past fifty years, since its main protagonist had lost his position of a "tsar" in Soviet biology and agriculture. The final part will outline the issues and questions arising from the history and historiography of "Lysenkoism," which are addressed in the present volumes, as well as possible directions for future studies.

The Lysenko Affair: A History

Trofim Denisovich Lysenko was undeniably a Soviet creature: his career as a scientist and state official was closely tied to major events in the history of his country and its science. Trofim was born on September 17, 1898, to a well-to-do peasant family in a small village in the Poltava region in the south-west of the Russian Empire (today's Ukraine). He attended an entry-level agricultural school in Poltava and would perhaps have become a prosperous farmer in his home village. But the 1917 Bolshevik Revolution created new opportunities for the peasant son. As soon as a civil war that had come on the tails of the Revolution spent its fury, in 1921, Trofim enrolled in "correspondence classes" at the Kiev Agricultural Institute, the leading agricultural school in the region. After graduation in 1925, he was sent as a rank-and-file researcher to a remote agricultural experimental station in Azerbaijan.

Lysenko soon came to head a small unit[6] that was assigned the task of introducing new varieties of beans, which could be sown in the winter and grow in the early spring and thus used as animal feed at the time when no local plants were available. This task prompted the young agronomist to investigate the role of temperature in plant development—the subject that would lead to his most celebrated "discovery" and provide a foundation for much of his subsequent theorizing. By exposing seeds and seedlings to different temperature regimes—a technique that was soon named "vernalization"—Lysenko was able to manipulate the timing of their germination and subsequent development in various agricultural plants. This research led him to formulate a concept of "phasic development," postulating that to be completed, different stages (phases) in the development of plants "require" specific temperatures.[7]

Lysenko's early work on vernalization attracted some notice of the scientific community, the press, and the agricultural officials. But it was a new revolution—Joseph Stalin's "revolution from above"[8]—that propelled Lysenko to the administrative heights of Soviet agricultural science. In the late 1920s, the launching of the ambitious first Five-Year Plan, crash industrialization, the forced collectivization of the peasantry, and extensive militarization (combined with a severe draught in the southern parts of the country) led to a horrific agricultural crisis and widespread famine, which prompted the Soviet government's massive investment in agricultural science. In early 1929, the Lenin All-Union Academy of Agricultural Sciences (VASKhNIL) was established as the center of research on all agricultural subjects and the country's leading plant scientist Nikolai Vavilov appointed as its president. The network of agricultural experimental stations and research institutes throughout the country was greatly expanded, with Vavilov's Institute of Plant Breeding in Leningrad serving as its "general staff."

The shortage of qualified personnel for these new institutions (coupled with a vicious attack on "old bourgeois specialists" opened by the 1928 infamous "Shakhty Trial")[9] boosted Lysenko's advancement up the academic ladder. In the fall of 1929, he was appointed head of a special laboratory created for his studies on vernalization at the newly established Ukrainian Institute of Genetics and [Plant] Breeding in Odessa. Initially, several Soviet plant scientists lent a hand to the growing popularity of Lysenko's ideas. Vavilov, for instance, excitedly reported on Lysenko's work to the VI International Genetics Congress held in Ithaca, in 1932. Vavilov saw in Lysenko's vernalization experiments a useful method that allowed manipulating the flowering of plants from different parts of the world he had collected during his numerous expeditions and thus made possible their crossings in order to breed new varieties of cultivated plants for Soviet agriculture. But Soviet agricultural officials began to actively promote the use of vernalization not as a laboratory procedure, but as an effective *agricultural technique* to increase the yield in a variety of plants, from beans and wheat to potatoes and cotton, all over the country. They also actively promoted the "discoverer" of this technique. In 1932, Lysenko acquired his own periodical, the *Bulletin of Vernalization*, which became an important channel for disseminating his ideas.[10] Two years later, he became a member of the Ukrainian Academy of Sciences and the "scientific principal" of the Odessa Institute. In 1935, Lysenko was appointed a member of VASKhNIL and a member of the Central Executive Committee, nominally the highest state office. The next year, he became the director of the Odessa Institute.

As Lysenko's institutional powers grew, so too did the scope of his theories. Originally, his concepts of vernalization and "phasic development" dealt exclusively with plant physiology and endeavored to explain the influence of temperature on the development of plants. In 1934, Isaak Prezent, a professional philosopher deeply involved in elaborating a "Marxist" interpretation of Darwinism, joined Lysenko's team.[11] It was Prezent who helped Lysenko devise a broad theoretical doctrine that connected vernalization with questions of heredity and evolution.[12] This doctrine was named "Michurinist biology" after Ivan Michurin (1855–1935), an amateur plant breeder hailed as a "Russian Luther Burbank" and accorded the status of a national hero in the Soviet Union in the early 1930s. Interchangeably, it was also named "Soviet creative Darwinism" to emphasize its connections to "Darwinism" that during the previous decade had been incorporated into the official Soviet ideology, Marxism.

The cornerstone of Michurinist biology was the notion of the "transformation of heredity" under the influence of external conditions. Thus, according to Lysenko, vernalization affected not only the physiology of plants but also their heredity, leading, for instance, to the hereditary change of "winter" varieties into "spring" ones. Ivan Michurin's favorite technique of plant grafting provided Lysenko with another way to "transform" heredity. Lysenko interpreted mutual influences of the scion and the rootstock in plant grafts as "vegetative hybridization" that perpetuated such influences in the heredity of offspring. Lysenko claimed that such vegetative hybrids display hereditary characteristics of both parental plants (species) and transfer them to the progeny in the same way sexual hybrids do.

Michurinist biology then openly contradicted the basic tenets of genetics, including Gregor Mendel's laws, Thomas Morgan's chromosomal theory, and the concept of the gene as a material unit of heredity, and supported the Lamarckian idea of the inheritance of the acquired characteristics. It clearly undermined the genetics-based principles of plant breeding, which underpinned the VASKhNIL vast program of creating new varieties of cultivated plants for Soviet agriculture. Not unexpectedly, it provoked severe criticism by several eminent plant breeders and geneticists. Opened on the pages of newspapers and periodicals, the polemic over Michurinist biology crested in two "public discussions" on "issues of genetics" initiated by Lysenko's opponents alarmed by his growing institutional reach and influence on the country's agricultural research, education, and policy.

The first such discussion took place in December 1936 at a special week-long session of VASKhNIL, marking the beginning of what Russian-American geneticist Theodosius Dobzhansky has fittingly named the *Lysenko Controversy*.[13] Leading geneticists and plant breeders, including Vavilov, Nikolai Kol'tsov, Petr Konstantinov, Petr Lisitsin, Alexander Serebrovskii, and H. J. Muller (at the time a senior geneticist at the USSR Academy of Sciences Institute of Genetics headed by Vavilov), advanced a detailed critique of the main theoretical postulates of Michurinist biology and Lysenko's experimental techniques.[14] In their turn, Lysenko and his disciples from the Odessa Institute launched a broad attack on genetics. Emphasizing the historical links between the development of genetics and eugenics (that had been banned in the Soviet Union in 1930), they employed the politicized discourse of the day to portray genetics as "foreign," "idealistic," "bourgeois," "racist," "fascist," "practically useless," and "anti-Darwinist" "Mendelism-Morganism" in contrast to their own "native," "materialistic," "proletarian," "socialist," "practically important," and "Darwinist" Michurinist biology. Geneticists defended their discipline by demonstrating the materialist character of its main theoretical concepts, such as Mendel's laws and Morgan's chromosomal theory. They also pointed to a broad antiracist and antifascist campaign mounted by leading Soviet and Anglo-American geneticists that was about to culminate in a special panel devoted to the critique of "racist and fascist perversions" of genetics at the forthcoming VII International Genetics Congress in Moscow in August 1937.

In some ways, the discussion helped geneticists fend off Lysenko's attack: VASKhNIL allocated considerable funds for expansion of genetics research. But it certainly undermined genetics and geneticists' authority among their patrons in the party-state apparatus: Lysenko's critique played an important role in the postponement of the VII International Genetics Congress by the highest party office—the Politburo. It also greatly alarmed the international genetics community, forcing the Congress's withdrawal from Moscow and its relocation to Edinburgh. Furthermore, the discussion failed to put a stop on Lysenko's further rise. In 1938, he was appointed president of VASKhNIL and a deputy head of the country's highest legislative body, the Supreme Soviet. A year later, he also became a member of the USSR Academy of Sciences and a member of its ruling council, the Presidium. Lysenko immediately employed his new administrative positions to promote his allies and supporters to key posts within the system of agricultural and biological institutions. In 1937–1939, Michurinist biology also began to enter textbooks on biology and plant breeding for secondary schools, agricultural colleges, and universities.

In October 1939, a second week-long discussion on "issues of genetics" took place at the Marx-Engels-Lenin Institute in Moscow. Triggered by a letter sent by a group of geneticists to Andrei Zhdanov, one of Stalin's lieutenants and a Secretary of the Central Committee of the Soviet Communist Party, the discussion was "adjudicated" by party philosophers, members of the editorial board of the country's major philosophy journal, *Under the Banner of Marxism.*[15] It was largely a replay of the first discussion, with a few minor changes in the rhetoric employed by both sides, which reflected recent drastic changes in the political situation. The August 1939 Molotov-Ribbentrop pact that had overnight made the Soviet Union a German ally forced Lysenko to drop his characterization of genetics as a "fascist science," while geneticists had to mute their references to the antifascist attitudes of their British and US colleagues. Lysenko and his disciples repeated their stock accusations against genetics as a pernicious "foreign," "idealistic," "bourgeois," "practically useless," and "anti-Darwinist" Mendelism-Morganism.[16] Geneticists, in turn, tried to refute the accusations by demonstrating certain practical achievements of their discipline and its adherence to the principles of Marxism and Darwinism, while repeating their critical assessments of Michurinist biology's experimental practices and theoretical conclusions.[17] The discussion ended in impasse. The party officials rebuked Lysenko for using his administrative powers to undermine the institutional positions of his opponents. But they also endorsed his critique of genetics as impractical and anti-Darwinist.[18]

The next year, the death of Kol'tsov and the arrest by the secret police of Vavilov and several of his coworkers opened the way for Lysenko to capture the last institutional bastions of genetics. He himself assumed the directorship of Vavilov's Institute of Genetics and secured the appointment of his faithful followers to the directorship of Vavilov's Institute of Plant Breeding and Kol'tsov's Institute of Experimental Biology. By the end of 1940, virtually all major centers of research in genetics and plant breeding within the system of VASKhNIL and the Academy of Sciences fell under the administrative control of Lysenko and his closest allies. Only a few laboratories and teaching departments at various universities remained in the hands of his opponents. The Nazi invasion of the Soviet Union in June 1941 plunged the country into World War II and relegated the controversy to the back burner.

As soon as the war ended, however, geneticists launched a broad campaign to restore the institutional basis of their discipline and to undermine Lysenko's influence on research, education, and policymaking in biology and agriculture. World War II profoundly altered every aspect of life in the country, including its science. The wartime scientific advances (such as the development of radar, antibiotics, computing machines, and new synthetic materials, to mention only a few iconic examples) greatly enhanced the authority of science, elevating several scientists, including plant geneticist Anton Zhebrak, to influential positions in the Soviet corridors of power. The Alliance of the Big Three—Great Britain, the Soviet Union, and the USA—restored the connections between the Soviet and Anglo-American scientific communities, which had been severed in the late 1930s. With a long history of close personal relations among its practitioners, in genetics, the revival of international contacts was particularly successful. As one would have expected, geneticists quickly capitalized on these developments. In 1945, Zhebrak initiated an intricate plot within the party-state apparatus and asked his British and US colleagues to open a "second front" against Lysenko.

Since the cancelation of the Moscow meetings of the VII International Genetics Congress in 1937, Western geneticists had observed in alarm Lysenko's growing influence on Soviet genetics and biology. They had followed closely the developing controversy between Lysenko and their Soviet colleagues over the basic principles of their discipline.[19] So when in summer 1945, they learned that their colleagues "personally and *confidentially* ask for support"[20] and that "the Soviet Government at the moment is definitely disposed toward giving considerable weight"[21] to the opinion of Western scientists, they enthusiastically responded to the request and launched broad critique of Lysenko's theories. As part of this well-coordinated campaign, Western geneticists issued an English translation of Lysenko's 1943 treatise on "Heredity and Its Variability" prepared by Dobzhansky and a thorough book-length analysis of Lysenko's research by British plant scientists.[22] They published more than a dozen of reviews and articles in scientific periodicals. It was within this critical campaign that the label "Lysenkoism" was coined and employed as a shorthand descriptor for Lysenko's doctrine.[23]

The anti-Lysenko publications in Britain and the USA did make a serious impression on the Soviet party-state officials. Soviet geneticists also capitalized on the recent synthesis of genetics and evolutionary theory accomplished by their Western colleagues to undermine Lysenko's claims that genetics was quintessentially "anti-Darwinist" while Michurinist biology exemplified "Soviet creative Darwinism." They prepared Russian translations of the key publications on the synthesis. Geneticists used the "second front" effectively to reverse the discouraging momentum of their prewar struggles with Lysenko and to justify their plans for the institutional expansion of their discipline. Against Lysenko's vocal objections, Nikolai Dubinin (a former student of Kol'tsov's) was elected to the USSR Academy of Sciences membership, and, together with Zhebrak, was working hard to establish a new Institute of Genetics under the academy's auspices.

What Soviet geneticists did not and could not plan for, however, was the Cold War. In the summer of 1947, with the shift of the Soviet Union's foreign policy from collaboration to confrontation with its wartime Allies, the elaborate Anglo-American links that had served Soviet geneticists so well suddenly became a dangerous liability. Lysenko and his allies immediately exploited this shift to discredit genetics' leading spokesmen, particularly Zhebrak and Dubinin,[24] and to stall geneticists' offensive. They also launched a broad press campaign to reaffirm the status of Lysenko's doctrine as "Soviet creative Darwinism." The next summer, with the Cold War reaching a crescendo over the status of divided Germany and its capital, Berlin, Lysenko managed to attract Stalin's personal attention to his struggle with geneticists and to secure the Soviet leader's personal support.

Lysenko's ultimate triumph over his opponents was staged as a "public discussion" at a special session of VASKhNIL, carefully planned and coordinated by the party-state officials, including Stalin himself. On July 31, 1948, Lysenko opened the session with a long speech "On the

Situation in Biological Science." The essence of Lysenko's address that three days later graced the front page of *Pravda*, the Soviet major newspaper, was a juxtaposition of two opposing trends in biology. One was pseudoscientific, idealist, scholastic, sterile, reactionary, anti-Darwinist Mendelism-Morganism; another scientific, materialist, creative, productive, progressive, Darwinist Michurinist biology. These two sets of antonymic labels clearly reflected the current sociopolitical situation: the escalating confrontation between "socialist East" and "capitalist West," or, as Lysenko phrased it, "two worlds—two ideologies in biology." Lysenko insisted that for a Soviet scientist, the only acceptable position was that of Michurinist biology.

The week-long meeting featured nearly fifty reports by Lysenko's supporters and only eight by his opponents. But even this meager opposition appeared unacceptable to the meeting organizers. On the last day, in his "concluding remarks," Lysenko announced that: "The Central Committee of the Communist Party has examined my report and approved it."[25] The next day, *Pravda* published Lysenko's remarks, making it clear to everyone concerned that Michurinist biology was now the officially sanctioned doctrine in the Soviet Union.

The VASKhNIL session inaugurated a massive propaganda campaign conducted under the slogan "for the undivided rule of Michurinist biology." Lysenko's address was immediately issued as a brochure with a print run of 300,000 (!) copies. By the end of August, the entire proceedings of the VASKhNIL session had been printed in 200,000 copies. During the fall of 1948, daily newspapers published countless articles by Lysenko's disciples that glorified their leader and his doctrine. The campaign quickly engulfed nearly all Soviet research and educational institutions in every scholarly field. Over the course of this campaign, Lysenko's opponents were demoted or fired, genetics laboratories were closed, genetics courses in agricultural schools and universities abolished, textbooks on Michurinist biology published, and Lysenko's supporters promoted to a variety of administrative positions within the country's entire system of biological, medical, pedagogical, and agricultural research and education. By the end of the year, the "undivided rule of Michurinist biology" had indeed been established.

The Michurinist campaign quickly spilled over the borders of the Soviet Union, becoming scientific cause celebre of the escalating Cold War. Lysenko's address and the entire proceedings of the August VASKhNIL session soon appeared in print not only in the newly born "people's democracies," but also in Afghanistan, Argentina, Austria, Britain, China, Egypt, France, Germany, Iran, Iraq, Japan, Lebanon, Turkey, the USA, and many other countries. The Soviet authorities often funded and always endorsed these publications, while a variety of local actors undertook their translation and dissemination.

As one would expect, the campaign gained particular momentum in the newborn socialist camp. In Bulgaria, East Germany, Czechoslovakia, Hungary, Poland, and Romania newspapers and magazines published editorials on the VASKhNIL meeting, "advances of Soviet science," and the tasks of local scientists "in light of Michurinist biology." Scientists immediately responded with countless articles, explaining what Michurinist biology was and how it should be practiced. Publications in the local press repeated ideological and political denunciations of genetics invented by Lysenko, emphasizing its alleged links with fascism and racism, its "practical sterility," and its contradictions to the sacred tenets of Marxism and Darwinism. Numerous meetings to discuss "the advances of Michurinist biology" supplemented the press campaign. As in the Soviet Union, genetics was condemned, Michurinist biology glorified, and individual biologists forced to comply with the "new genetics."

The wide campaign in the Eastern bloc was matched by a counter campaign in the West. Major newspapers in practically every Western country published editorials and reports on the subject. Many newspapers and magazines also published commentaries by leading local biologists and geneticists. To give but one example, from August 13 to December 20, 1948, the *New York Times* carried no fewer than two dozen items on the subject. Meanwhile, the BBC and the Voice of America broadcast special programs on Lysenko and his doctrine. As in the East, various interest groups held meetings to discuss Michurinist biology in France, Italy, Belgium, England, Holland, Sweden, and the USA. Often, local communist organizations spearheaded the meetings, but other political and professional groups, ranging from the Engels Society to the American Genetics Society to the Federation of Atomic Scientists, also hosted such gatherings. Societies of the "Friends of Michurin" sprung up in Belgium, France, Italy, and Japan.

The overriding tenor of this campaign was the exact reversal of Lysenko's own condemnations of Mendelism–Morganism. Just as Lysenko portrayed genetics as an "American," "imperialist," "racist," and "fascist" pseudoscience, many Western commentators presented Michurinist biology as a "Soviet," "Communist," "Marxist," and "totalitarian" pseudoscience.[26] It was in the course of this broad campaign that the label

"Lysenkoism" began to signify not merely Lysenko's doctrine, but the infringement of Soviet political authorities on academic freedom, the corrupting influence of Marxist ideology on scientific culture, and pseudoscience writ large. The anti-Lysenko campaign became an integral part of the Cold War anticommunist propaganda.

Although many Western observers mourned the "death of genetics" in the Soviet Union, Lysenko's hegemony did not last very long.[27] Ironically, the very same Cold War that had driven the officially sanctioned banishment of genetics in the Soviet Union made possible its "revival" just a few years later. The newly discovered importance of genetics in the age of atomic weapons and space exploration helped Soviet geneticists build a new institutional basis outside of Lysenko's administrative reach, under the protective umbrella of physics and chemistry research establishments involved in the Soviet nuclear and space programs. Stalin's death in March 1953 and subsequent "de-Stalinization" initiated by his successor Nikita Khrushchev greatly facilitated their efforts.

Soviet geneticists had learned important lessons from their previous battles with Lysenko and his allies. This time, they did not want to engage in any "public discussions" over "issues of genetics." They wanted a decision by the party-state apparatus that would dismantle Lysenko's monopoly. In the early fall of 1953, biologists and agronomists began bombarding party officials with letters and manuscripts detailing Lysenko's negative role in the development of their fields. This letter-writing campaign reached a peak in October 1955, when 297 scientists (among them leading Soviet physicists, chemists, biologists, and mathematicians) signed a petition to the Central Committee that described the detrimental effects of Lysenko's monopoly over Soviet biology and agriculture.

The campaign proved only partially successful. In 1956, Lysenko was fired from the VASKhNIL presidency, but retained his membership in the Presidium of the Academy of Sciences and his directorship of the academy's Institute of Genetics. Nevertheless, two years later, geneticists succeeded in creating a new Institute of Genetics under Dubinin's directorship, as part of *Akademgorodok*, a new "science city" built by the Academy of Sciences in Novosibirsk.[28] In the wake of Khrushchev's campaign for "mastering the virgin lands" (aimed at developing vast tracts of land in Siberia and Kazakhstan for agricultural use), however, Lysenko managed to gain the ear of the party leader. Khrushchev even reinstated him (briefly, from August 1961 to April 1962) at the helm of VASKhNIL. With the ousting of Khrushchev in October 1964, Lysenko's career came to an end as well.

In early 1965, Lysenko was discharged from all administrative positions. His journal, *Agrobiology*, was discontinued. He was allowed to retain only a small laboratory at the "Lenin Hills" estate near Moscow. Lysenko continued to enjoy the privileges accorded by his status as a full member of both VASKhNIL and the USSR Academy of Sciences, and occasionally, even showed up at the annual meetings of these academies and pestered party-state officials with complaints about his "persecution" by geneticists. Otherwise, he virtually disappeared from the public eye. On November 20, 1976, Lysenko died in the Kremlin hospital and was quietly buried at the Kuntsevo Cemetery reserved for eminent writers, composers, artists, and scientists. With the death of its main protagonist, the Lysenko affair came to an end.

But the controversy over his life and work continued.

The Lysenko Controversy: A Historiography

As soon as Lysenko lost administrative control over Soviet agriculture and biology, first attempts at a historical assessment of his life and works began in both the Soviet Union and the West.[29] These efforts aimed at answering two fundamental, interrelated questions: What could explain Lysenko's rise and eventual fall? And why was genetics banished and replaced by Michurinist biology in the Soviet Union in 1948, but "restored" in the 1950s?

In the Soviet Union, the first assessments of the Lysenko episode took place within the context of Khrushchev's "de-Stalinization" campaign, and clearly reflected its two main directions: "to overcome the negative consequences of Stalin's personality cult" and "to rehabilitate victims of Stalinist repression," as Khrushchev himself characterized them in his famous "secret speech" to the Twentieth Party Congress in February 1956. As a result of Khrushchev's "Thaw," a number of scientists who had been arrested, imprisoned, and/or executed during the Stalin era were "rehabilitated." One such scientist was Nikolai Vavilov, Lysenko's most famous opponent, who had been arrested in 1940, and three years later, died in prison from starvation.[30] Biographies of Vavilov published in the mid-1960s by professional writers Mark Popovskii and Semen Reznik described (however briefly) Vavilov's polemic with Lysenko and linked, though indirectly, Vavilov's arrest and tragic death to his opposition to Lysenko's theories.[31] Around the same time, biologist Zhores Medvedev produced a detailed account of Lysenko's "rise and fall." Medvedev's book portrayed Lysenko's ascent as a direct consequence of Stalin's "personality cult," which led to the elevation of a poorly educated man to the highest administrative positions in Soviet science and his "outdated" theories to the status of "official" biology. Medvedev explained Lysenko's eventual decline along the lines of Khrushchev's "de-Stalinization" rhetoric as a "correction of past mistakes" by the Communist Party and its leadership that restored "true Leninist policies" and returned genetics to its rightful place in the Soviet science system. But even such a "politically correct" analysis appeared unacceptable to the Soviet authorities. Censorship forbade the publication of Medvedev's book. Its manuscript was circulated only clandestinely in *samizdat* and its author was briefly confined to a mental hospital as punishment.[32]

The banishment of Medvedev's book made Lysenko's story a virtual taboo in the Soviet Union. Unable to openly address the Lysenko controversy, biologists, philosophers, and historians of science adopted a different tactic in their efforts to "rehabilitate" genetics. They found a way around the official wall of silence by publishing historical accounts that rarely—if at all—even mentioned Lysenko's name, but aimed at undermining the basic principles of Lysenko's Michurinist biology and dispelling his portrayal of genetics as "foreign," "idealist," "anti-Darwinist" Mendelism-Morganism. Several works detailed "Russian contributions" to genetics, demonstrating its deep "national" roots.[33] Others were devoted to disassociating Michurin from Lysenko by proving Michurin's favorable attitude toward genetics.[34] Still others provided historical analyses of the Lamarckian idea of the inheritance of the acquired characteristics, showing how genetics had disproved this "outdated" concept.[35] Several works demonstrated that, contrary to Lysenko's claims, genetics was not only fully compatible with Marxism, but in fact provided one of the best examples of the actual application of dialectical materialism in science.[36] But the liberalization of Soviet life could only go so far: with the end of the "Thaw," even such oblique analyses of the Lysenko controversy virtually disappeared. From the early 1970s on, works on the history of Soviet genetics, biology, and science more generally contained only veiled references to Lysenko.[37]

In the West, the situation was quite different. During the wide anti-Lysenko campaign of the late 1940s and 1950s, its participants found simple answers to the basic questions raised by the Lysenko controversy. They saw the rise of Lysenko as a direct result of the totalitarian system that exerted complete control over all facets of Soviet life, including science.[38] They discovered the chief reason for "why the Communists destroyed genetics" in "the fact that it was incompatible with the biological doctrines of Marx and Engels," while Lysenko's doctrine represented "a special Communist biology" developed under the direct influence of Marxism.[39]

Although continuing to employ the totalitarian model of Soviet society and its history, many of the new accounts of the Lysenko controversy that began to appear in the 1960s and 1970s challenged the perceived view of its intimate connections with Soviet ideology and Marxism. US historian David Joravsky carefully analyzed Lysenko's career and his struggles against Soviet genetics and geneticists. He argued forcefully that in the Lysenko affair, "farming was the basic problem, not theoretical ideology."[40] On the other hand, the revival of Marxism in Western academia in the late 1960s spurred numerous attempts to distance Marxism as a philosophical doctrine from Stalinism and thus from its most egregious manifestation in science, Lysenkoism. In his voluminous examination of the important role that Marxist dialectical materialism had played in the development of Soviet science, including physics, chemistry, and biology, US historian of science Loren R. Graham explicitly denied any links between Marxism and Lysenko's doctrine. Even though he provided a detailed outline of the controversy, he dismissed it as "a chapter in the history of pseudo-science rather than the history of science."[41] In the late 1970s, a number of British, French, Italian, and US biologists, historians, and philosophers advanced similar views and debated the exact place of Marxism in the Lysenko affair.[42] This broad debate also enticed several historians to look more closely at the role of several Western biologists who considered themselves Marxists, particularly J. B. S. Haldane and J. D. Bernal, to begin examining the ways the Lysenko controversy had been perceived in the West during the 1950s.[43]

Nevertheless, the portrayal of the Lysenko controversy as an example of science corrupted or perverted by political and/or ideological authorities continued to enjoy wide popularity,[44] facilitating further expansion of the uses of the term "Lysenkoism" beyond its original Soviet contexts.[45] It even spanned a new label, "Neo-Lysenkoism," coined in 1983 by Harvard geneticist Bernard D. Davis in his critique of Stephen Jay Gould's book *The Mismeasure of Man*.[46] This label stretched the meaning of Lysenkoism even further to cover a wide variety of politically, ideologically, and financially motivated ideas, individuals, and organizations in public debates about science, politics, pedagogy, religion, medicine, and economics. Continuing this trend, a few years later, "Club de l'Horloge," a French arch-conservative group, established a special "Lysenko Prize." This "anti-prize" was (and still is) to be "awarded annually to an author or a personality who, through his writings or his actions, made an exemplary contribution to [the public] misinformation in scientific or historical matters, with the use of ideological methods and arguments."[47] In a peculiar twist of its founders' political schemata, the first "Lysenko prize" was awarded in 1990 to Albert Jacquard, a French population geneticist renowned for his active opposition to racist interpretations of human genetic diversity.[48]

Yet the simplistic dismissal of the Lysenko controversy as an instance of pseudoscience endorsed by political and/or ideological authorities did not go unchallenged. Employing the new approaches of social and cultural history, some Western scholars began to investigate not the political and/or ideological, but the intellectual, institutional, and disciplinary dimensions of the controversy. US historian of science Mark B. Adams produced several detailed studies of the institutional development of Soviet genetics, which demonstrated the critical role of scientific networks in the revival of Soviet genetics after Stalin's death.[49] He also wrote a short scientific biography of Lysenko and was instrumental in bringing to English readers several articles on the subject written by Soviet historians of science, which at the time could not be published in the USSR.[50] Norwegian historian of science Nils Roll-Hunsen carefully examined Lysenko's early works on vernalization and came to the conclusion that, far from representing a "pseudoscience," they in fact fell squarely within the accepted norms of contemporary research in plant physiology and had indeed inspired a variety of follow-up studies by scientists in Great Britain, the USA, and elsewhere.[51] US historian of science Douglas R. Weiner examined the long tradition of transformist biology and the wide popularity of J. B. Lamarck's ideas about the inheritance of the acquired characteristics in Russia as important intellectual contexts that facilitated the rise and acceptance of Lysenko's doctrine.[52] Canadian historian of science Jan Sapp suggested that in his struggles against Mendelism-Morganism, Lysenko had been able to exploit the then poorly understood instances of cytoplasmic heredity, which apparently defied Mendel's laws and contradicted Morgan's concept of chromosomal heredity.[53] The inaccessibility of Soviet archives, however, forced Western scholars to rely largely on published sources and memoirs, substantially restricting the range of questions they could address and lending a hypothetical character to some of their interpretations.

The dawn of Mikhail Gorbachev's *perestroika* and *glasnost'* in the mid-1980s profoundly affected historical examinations of the Lysenko controversy in both the USSR and the West. In the Soviet Union, "rethinking of the country's past" and "filling up gaps" became the main direction of all historical scholarship. In the history of Soviet biology, one such obvious gap was the Lysenko controversy. As during Khrushchev's Thaw, the first to respond to the call of the times were professional writers. The publication of two novels, Vladimir Dudintsev's *White Robes* (in 1986) and Daniil Granin's *The Bison* (in 1987), both of which presented fictionalized accounts of the controversy, enticed wide public discussions, especially within the communities of biologists and historians of science.[54] They spearheaded the publication of numerous memoirs and documents, including nearly all of the manuscripts and letters with critical assessment of Lysenko's role in Soviet biology and agriculture, which had been sent to the Central Committee during the 1940s and early 1950s.[55]

In 1991, the end of the Cold War and the dissolution of the Soviet Union opened up the previously inaccessible archives of the party-state agencies and the country's borders for both Russian and Western historians. These developments brought to light enormous amounts of new materials, including, for the first time, the proofs of Stalin's personal involvement in the later stages of the Lysenko affair, and facilitated a "convergence" of the Russian and Western scholarship on Lysenko and Lysenkoism. This convergence took the form of two distinct genres that reflected two major historiographic traditions by that time well-entrenched in the Lysenko scholarship: the "oppressed science" and the "socio-cultural history of science."

Works written in the genre of "oppressed science" deployed the new archival materials to continue the tradition (established during the post 1948 anti-Lysenko campaign in the West) to draw a sharp and fast divide between Soviet science and the "totalitarian" Soviet state, reducing the multitude of interactions between science and society to "oppression" and "resistance."[56] The entire history of Soviet science was portrayed in simple black-and-white dichotomies of "executioners" and "victims," "triumphs" and "tragedies," "heroes" and "villains," loaded with explicit moral judgments of historical events and actors. And, of course, the Lysenko affair provided a prime example of such oppression and resistance, not only in the Soviet Union, but also in the entire socialist camp.[57]

Works written in the second genre used (often the same) new archival materials to apply to the Lysenko controversy the sophisticated social and cultural analysis advanced in the history of Western (and occasionally Soviet) science during the 1970s and 1980s. These works rejected simplistic dichotomies and called for a careful examination of the multiple ways in which science and scientists interacted with the Soviet state and society.[58] They supplemented the descriptions of oppression and resistance with analyses of negotiations, borrowings, adaptations, overlapping interests, cooptation, symbiotic relations, appropriations, cooperation, shared language, rituals, and rhetoric, mutual influences, disciplinary and institutional competition, patron–client relations, symbolic exchanges, and numerous other forms of interactions.[59] As one might expect, the Lysenko controversy became a focal point of such analyses.[60]

During the 1990s, new works produced largely by Russian historians challenged the previous accounts in several important ways.[61] First of all, using new archival materials, they demonstrated that, far from being a conflict between the monolithic "totalitarian" state and an equally monolithic "autonomous science," the Lysenko controversy reflected a particular symbiotic relationship between the party-state apparatus and the Soviet scientific community and the ability of different competing groups (e.g., geneticists and Lysenko's team) to exploit this relationship to their own advantage. Thus, contrary to the prevalent depiction of the controversy as an interrupted series of Lysenko's assaults on[62] and victories over genetics, culminating in the August 1948 VASKhNIL meeting, they underscored the decisive role of changing priorities in the party-state apparatus, following the complex mechanics and dynamics of the controversy and the changing fortunes of its main protagonists (i.e., the rise and fall of Lysenko versus the rise and fall and rise again of Soviet genetics).

Second, they demonstrated that Marxism did play an important role in the controversy, but not as an "ideological bludgeon" used by Lysenko to pound his opponents into submission. Rather, Marxism provided the vocabulary for a new language shared by *all* the participants in the Soviet science system. This Newspeak served an important function of translating, on the one hand, scientists' specific (and often quite esoteric) interests into the language understood and appreciated by their patrons and partners in the party-state apparatus, and on the other hand, current (often shifting and contradictory) priorities of party-state bureaucrats into the language understood and appreciated by scientists. The new works showed that already in the early 1920s, Darwinism had become an important part of such Newspeak developed by Soviet biologists, which explained the important role that the struggle for authority over evolutionary issues played in the Lysenko controversy at every stage in its development—from the early 1930s through the late 1960s.

Last, but not least, they dispelled the traditional perception of the controversy as an exclusively "Soviet" affair driven by internal issues, actors, and events, demonstrating the important role that events and actors outside of the Soviet Union played at its various stages.[63] In particular, they emphasized the critical role of the concurrent international political situation (from the rise of Nazism in Germany through World War II to the Cold War) in shaping the dynamic of the controversy and eventually turning it into a global phenomenon that spread over Europe, Asia, and the Americas. They called for a careful examination of converging efforts of various Soviet agents and agencies to "export" and numerous local agents and agencies to "import" (or to block the spread of) Michurinist biology, as well as different interests and agendas that underpinned their actions on both sides of the Iron Curtain.[64]

The first decade of the new millennium saw a virtual explosion of historical studies on Lysenko and Lysenkoism, which examined the specifics of the Lysenko controversy not only in the Soviet Union but also in Belgium, Britain, China, France, East and West Germany, Holland, Italy, Japan, Poland, and the USA. Some of these studies, especially among those produced in the countries of the former "Soviet bloc," continued along the lines of the older "totalitarian" interpretations.[65] Many others followed the newer "socio-cultural" approaches.[66] These works indicated that in each individual country, pre-existing cultural and scientific traditions, institutional structures of local scientific communities, particularities of local power struggles, and political engagements of individuals shaped directions, forms, outcomes, durations, targets, means, and audiences of the pro- and contra-Lysenko campaigns, while after 1948, the Cold War provided overriding political, economic, and ideological contexts for the events. Many scholars felt that the time has come to look beyond the particularities of individual cases to discern certain general patterns and common themes, to develop cohesive analytical and explanatory frameworks, and to take a comparative approach in their analysis and understanding of the global "Lysenko phenomenon."

Answering the call of the times, William deJong-Lambert, who had conducted extensive studies of the Michurinist campaign in Poland and the anti-Lysenko campaign waged by US and British biologists, single-handedly organized an international conference on "Lysenkoism" in December 2009 in New York City. The conference brought together some thirty historians from Canada, Czech Republic, Denmark, Germany, Italy, Japan, Mexico, Norway, Russia, and the USA, as well as a large audience from New York City's various universities. For two days, it provided a splendid forum for presentations, debates, and informal discussions.[67]

Although some of the presentations continued the tradition of the "oppressed science," the conference as a whole focused largely on the analysis of the pro/contra Lysenko campaigns spurred by the August 1948 VASKhNIL meeting as a symbol, an instrument, and a focal point of the Cold War confrontation between the two competing blocs, East and West. It convincingly demonstrated the sheer size, longevity, and geographical reach of the Lysenko controversy that had spanned several decades and enveloped almost the entire globe. Along with details specific to each individual case, the conference participants detected a host of similar features the controversy had assumed in different locales, suggesting that all the actors had deployed the controversy as a particular *cultural resource* to address a variety of (often the same) issues individuals and groups in different countries faced in the domestic and international arenas.

As several presenters convincingly demonstrated, for government officials and politicians in such diverse countries as China, Japan, and Poland, the controversy had offered an easy way to demonstrate their adherence to a particular side in the Cold War confrontation and to showcase their loyalty, exploiting the economic, military, political, and any other benefits such loyalty could entail in a world polarized by the Cold War.[68] Western biologists (and geneticists in particular) had used the controversy to debate and eventually reach (at least provisionally) a consensus regarding such burning questions of their discipline as the inheritance of the acquired characteristics, cytoplasmic heredity, gene theory, and mechanisms of biological evolution. The controversy also provided them with a useful tool to shape the public image of their discipline (and science more generally),[69] and proved instrumental in institution building in the biological sciences on both sides of the Iron Curtain, not only in war-ridden Europe, but also in Asia and South America.

Other presentations showed that in the West, the controversy had reached far beyond genetics and biology and had served a purpose far larger than exposing the fallacies of the Soviets, forging intra-disciplinary consensus, or boosting the institutional and disciplinary standing of genetics. Scientists from many other fields, ranging from physics and mathematics to chemistry and psychology, had eagerly joined in the fray to discuss the relations of science to the state, industry, ideology, and society writ large. For many Western scientists, Lysenko had become a convenient straw man in heated debates over the exact role that state agents and agencies could and should play in science, and over particular forms of state-funded "big" science that had come into being during World War II. These debates revolved around the issues of "academic freedom" and contrasted the two alternative models of "big" science: "totalitarian" (read "state-run") embodied by Soviet Michurinist biology and "democratic" (read "scientist-run") embodied by Western Mendelian genetics. The anti-Lysenko campaign thus had been waged to educate state officials, scientists, and the public regarding the new relationship between science and the state, which had emerged during World War II and been cemented in the Cold War. It had served as an important argument in the lengthy negotiations between scientists and their patrons over the principles of operation of such science-funding agencies as the National Science Foundation, the National Institutes of Health, the National Endowment for the Humanities, and the Atomic Energy Commission in the USA, the National Research Council and the Medical Research Council in Britain, and the Centre national de la recherche scientifique (CNRS) in France. For instance, US scientists had regularly referred to Lysenko's triumph in the Soviet Union to defend their own independence from the state's administrative interventions and ideological pressures,[70] routinely applying the label "Lysenkoism" to attempts to destroy "academic freedom" not just in the Soviet Union, but everywhere.

Several contributions indicated that the Lysenko controversy had proved influential not only in debates over the "external" relations of science. It had also served as a whetstone in impassioned arguments over what science—and by extension, pseudoscience—is.[71] The controversy had provided a template for answering a series of critical questions about the nature of science itself: What constitutes appropriate/inappropriate and legitimate/illegitimate scientific practices in terms of both investigative practices that generate scientific knowledge and social practices that generate careers, patrons, and institutions? What makes an individual a scientist? How should scientific bodies be run and scientific controversies resolved? Could a philosophy—be it Marxist, positivist, or any other—actually direct scientific research? The conference presentations and discussions suggested that answers to these questions had been essential in shaping individual and group identities—including political affiliations (e.g., membership in the Communist and other left-wing parties), public responsibilities, and scientific loyalties—of numerous scientists in various countries, ranging from Poland and Czechoslovakia to Italy and the USA.[72]

The conference as a whole convincingly showed that over the course of the last fifty years, our understanding of both the *Lysenko Affair* in the Soviet Union and the global *Lysenko Controversy* has progressed tremendously. It helped set agendas for further research and stimulated a large number of new studies. It fostered the formation of an international community of scholars that, in the summer of 2012, held a follow-up conference on "Lysenkoism" in Vienna, which provided a foundation for the present volumes.

Questions and Issues: Affair, Controversy, and Effects

Although decentering Lysenkoism from the zone of Russian-Soviet science was the main motif of the Vienna conference, a re-examination of what we thought we knew about what took place in its namesake's homeland constituted one of its major themes. That much of this history was oversimplified is unsurprising. However, some of the chapters which follow show that questions such as why innovations similar to Lysenko's celebrated vernalization were *not* adopted, whether the Soviet press can really be blamed for promoting Lysenko despite a more cautious reception from the scientific establishment, as well as whether or not good work occurred despite Lysenko's dominance, are still to be addressed. There is also the question of whether or not the term "Lysenkoist" is actually far more slippery than is reflected in the historiography.

Olga Elina asks us to think about the question of why Lysenko's work was promoted, and others were not, in her chapter on Nikolai and Boris Demchinsky, the father-and-son team who developed the "bed cultivation" technique. The reputed success of this method was more or less the same as that of Lysenko's vernalization. Yet the Demchinskys' panacea for chronic wheat shortages never received the same level of support as Lysenko's, and they dropped into obscurity even as the latter rose to fame and notoriety. In chronicling the timeline of these events, Elina examines the mechanisms that account for one method being promoted while another is neglected, as well as the motives of panacea promoters who, purportedly, are interested in nothing more than improving agricultural production.

The traditional story, as told by authors such as Zhores Medvedev and David Joravsky, is that feature stories in the press launched Lysenko to scientific celebrity in the Soviet Union. By comprehensively examining the trail of correspondence between officials at the Ukrainian Ministry of

Agriculture and taking into account the unsophisticated state of science journalism in the USSR at the time, Lukas Joos deconstructs this version of events and places full blame upon the scientific establishment. Joos's chapter refocuses our attention on the extent to which factors presumably beyond the control of scientists themselves have unfairly carried the onus of being responsible for the rise of Lysenko. It also reminds us that the role of scientists in promoting Lysenko needs to be reconsidered.

Mark Tauger also problematizes the assumptions about Lysenko's career in his chapter on Pavel Luk'ianenko, a successful plant breeder who conducted valuable research under Lysenko. Tauger raises the important question of whether Lysenko's control and influence were as absolute as has been portrayed. As Tauger writes, Luk'ianenko independently developed a high-yielding semi-dwarf strain of wheat, Bezostaia-1, concurrent with the man credited with founding the "Green Revolution," US geneticist Norman Borlaug. This indicates that progress in Soviet plant genetics was by no means halted during the period of Lysenko's greatest influence. Of perhaps equal importance is that Borlaug acknowledged Bezostaia-1 as being as valuable as the varieties bred by his research group. Once again, earlier perceptions that capitalist and communist agriculture operated along mutually exclusive lines during Lysenko's tenure are called into question.

The question of whether the label "Lysenkoist" actually means anything is explored by Pat Simpson in her chapter on Alexander Kots, the founder and director of the Darwin Museum in Moscow. Simpson's account of how easily Kots manipulated the exhibits inside his museum in response to events in Soviet biology raises the question of how one would define culpability. Was Kots a Lysenko supporter, or a canny curator? Simpson also opens up broad new avenues for research of the impact of Lysenkoism upon art, literature, music, film, online video, as well as any number of other aesthetic formats. Although a copy of the 1975 NOVA television documentary on Lysenko has, thus far, proved elusive, there must be other areas where we spot the influence of the controversy embedded in heady concepts such as the transformation of nature and the nature of science itself.

Although Lysenko's name certainly appeared in the foreign press prior to August 1948, the controversy only began once his influence spread and the West responded. Though it was once assumed that scientists in Soviet-allied states marched along according to what was dictated from Moscow, recent scholarship has opened up a fascinating diversity of outcomes, depending upon when and where you look. By the same token, the anti-Lysenko campaign was once perceived as a united front, with only a few random outliers and dead-enders reluctant to throw in full support.

We now know that the terrain on the other side was fraught in its own ways, with geneticists in the West politicking to secure their positions.

Although as recently as a few decades ago, historians would likely have been hard-pressed to come up with any kind of positive impact which resulted from Lysenko's anti-genetics campaign, Miklós Müller and Gábor Pallo's account shows us that it led to—for the first time—the institutionalization of biology in Hungary. Müller and Pallo also outline five strategies according to which Hungarian biologists responded to Lysenkoism and conclude that the best option was active opportunism. Although apparently complying with the new Soviet biology, active opportunists were positioned to defend research they regarded as important and protect their junior colleagues doing good work. In this way, they preserved and promoted the long-term development of Hungarian biology. An additional irony uncovered in this account is that it was Lysenko himself, thanks to a disastrous performance at the Hungarian Academy of Sciences in 1960, who contributed most to the downfall of Lysenkoism in Hungary (Fig. 1).

Fig. 1 Lysenko delivering a speech at the Hungarian Academy of Sciences during his 1960 visit to Budapest
Source: Mrs. Diana Hay, Hungarian Academy of Sciences

Francesco Cassata provides us with another positive outcome of Lysenkoism, showing how it served as a cultural resource for Italian geneticists seeking to consolidate their discipline and integrate it into the international community. Among their primary challenges was the recent close association between genetics and eugenics in fascist Italy, as well as geneticists' rivalry with the two other primary disciplinary groups in Italian biology, plant breeders and medical geneticists. This effort, facilitated by the organization of the IX International Congress of Genetics in Bellagio, Italy in 1953, set the course for the progress of Italian genetics in the following decades.

As surprising as what we might tentatively categorize as the "ameliorative" impact of Lysenkoism, equally intriguing is what happened in Romania, as described by Cristiana Oghina-Pavie. Although historians such as Nils Roll-Hansen have long raised questions as to what extent we could classify Lysenko's theories as "Lamarckian," Oghina-Pavie complicates this picture by showing how the association between Neo-Lamarckism and Lysenkoism extended the life of the latter long past its due date in the other "little democracies" of the region. This longevity was the product of Romanians' long fascination with every aspect of French science and culture. That elements of Lysenkoism still appeared in Romanian biology curriculae into the 1970s destabilizes the Iron Curtain and the Moscow hand narratives. As Oghina-Pavie shows, while the Cold War may have been essential to Lysenko gaining Stalin's support, what happened next was influenced by local factors in both East and West, which often played a role independent of the rivalry between the USA and the USSR.

It is also now clear that the fault lines of this rivalry fractured the relationships between those who had once seemed united on other causes. In his chapter, William deJong-Lambert describes how Nobel Prize-winning geneticist Hermann J. Muller's efforts to hide his attempt to persuade Stalin to implement his eugenic plan during the interwar period, as well as his worries that this contributed to the cancellation of the VII International Genetics Congress set for Moscow, led him to deceive British geneticist J. B. S. Haldane about Lysenko's rising influence in Soviet biology. Although Haldane's credentials as a prominent member of the community of "Marxist biologists" in Britain are unquestioned, it is hard to imagine he would have endorsed Lysenko as he did, had Muller told him the truth. Once again, the dichotomies of honesty versus deception, the way scientists behaved in the West versus the Soviet Bloc, are called into question, providing a far more nuanced portrait than historians painted for most of the controversy's history.

Indeed, it is now clear that the determination of Western geneticists to portray Lysenkoism as a cause that one must be either for or against had numerous negative outcomes, which are only recently being explored. Hirofumi Saito's account of the situation in postwar Japan not only highlights the significance of local differences, it also shows the negative impact of the anti-Lysenko campaign in the USA upon the development of Japanese genetics. In this instance, it was the USA, not the USSR, which played an effectively censorial role. As Saito describes, after World War II, Japanese geneticists were very eager to learn more about the developments in their discipline which had taken place while the country had effectively been cut off from the mainstream of biological science. Not only did they not discriminate between knowledge based in either the USA or the Soviet Union, they were deeply interested in the influence of the environment in phenotypic expression. This drew them to Lysenko's work. The US domination over every aspect of Japanese science, politics, and culture during this period made such avenues of inquiry impossible to pursue, thus snuffing out what may have been a thriving field of hereditary research.

Luis Campos also widens our perspective on what areas of biological research were eliminated thanks to the campaigns for, and against, Lysenko. In this case, he locates chromosomal mutations, a growing field of inquiry in both the USA and the USSR by the 1930s. As Campos documents, US geneticists were reluctant to consider other, nongenic levels of heredity and mutation, out of fear of being associated with Lysenkoism. Chromosomal mutations, therefore, can be considered as an important, heretofore unacknowledged casualty of Lysenkoism.

This leads us to the question of what Lysenkoism actually means, addressed by John Marks in his chapter on whether there is anything to be learned from the episode. Was it an anomaly, or an example of what happens when science is infiltrated on some level by elements alien to scientific methodology and practice? As Marks argues, the way the term "Lysenkoism" is used falsely implies the existence of an objective science isolated from the culture that produces it. Marks also affirms the evidence assembled in the previous chapters, that anti-Lysenkoism must be considered in terms of the distortions it produced, and continues to produce, of the same nature as that which it was attacking.

That such issues are not only of historical interest is affirmed by Eduard Kolchinsky, in his chapter on the status of Lysenko's reputation in Russia today. Kolchinsky takes us full circle to the present, describing how the relationship between politics, society, and religion in Russia has fostered a revival of Lysenkoism. This points to Lysenko's lingering influence upon Russian scientific history and his reification as a model in Russian culture. That politics and religion also structure the place of science in contemporary societies worldwide as well goes without saying. This points to another issue: The fact that the term "Lysenkoism" has become a synonym for "pseudoscience."

We know, etymologically speaking, that attaching the suffix "ism" to a word is a way of criticizing it by implying that there is something false about what it refers to. The point is to trivialize and signal that the content attached to the signifier not be taken seriously. This would seem good cause to abandon the term "Lysenkoism" and search for an alternative, or simply be more judicious with our language use when discussing the topic. However, we embrace the term as a critical lens to examine the motives of the actors who enlist it in any scientific controversy, whether it be climate change, evolution, acupuncture, or any number of other topics to which it has been—and will continue to be—applied. The advantage is that those who employ it do so to project a negative image upon their opponent(s), but their use of the term actually serves to expose them and enables us to better understand the assumptions they harbor and beliefs they would impose.

Finally, we have the question of what research remains to be done, and for better or worse, the answer is a lot. We have yet to hear anything about how Lysenko affected biological science in the far-flung reaches of the Soviet Empire. Presumably Lysenko's anti-genetics campaign influenced major agricultural initiatives in the USSR, such as cotton production in Uzbekistan, which, at least in the short term, resulted in major environmental disasters, such as the draining of the Aral Sea. There is also evidence that African mammals were introduced into colder regions of the Soviet Union as a feature of Lysenkoist experiments in animal breeding, but so far we have little detail of who initiated them and what the results were. Following the popular trope of "heroes and villains," historical examinations have focused on Lysenko himself and his most prominent opponents, such as Vavilov and Kol'tsov, while the role of Lysenko's closest collaborators and promoters remains largely unexamined. We need to know much more about Lysenko's support base and networks among the party-state officials, the scientific community, and rank-and-file agricultural practitioners.

The geographical reach of the Lysenko controversy remains a major subject for future research. We know of a hot debate between Lysenko critic Julian Huxley and Soviet Lysenkoist Nikolai Nuzhdin in Karachi, Pakistan, in 1954, but what was the impression Pakistani biologists came away with? Where did the outcome of this debate go in terms of its influence upon their careers and the development of biology in Pakistan in the following decades? When J. B. S. Haldane emigrated to India in 1957, bringing his reputation as an advocate of Lysenko with him, how did this affect the research and ideas of those who worked with him? What about Central and South America, and the Caribbean—particularly Cuba? What about Africa or other parts of Asia besides China and Japan? Such case studies would be extremely illuminating in terms of deepening our knowledge of the topic and would also constitute important contributions to the history of science in whatever region they covered. The present volumes lay the groundwork for such research and will, hopefully, even inspire it.

Notes

1. S. J. Gould, *Hen's Teeth and Horse's Toes* (New York: Norton, 1983), p. 135.
2. See, for instance, P. F. Kononkov, *Dva mira—dve ideologii. O polozhenii v biologicheskikh i sel'sko-khoziaistvennykh naukakh v Rossii v sovetskii i post-sovetskii period* (Moscow: OOO Luch, 2014).
3. This was how US botanist Leo Kartman, who apparently coined the label, used it. See L. Kartman, "Soviet Genetics and the 'Autonomy of Science'," *Scientific Monthly*, 1945, 61(1): 67–70.
4. See, for instance, M. W. Mikulak, "Trofim Denisovich Lysenko," in G. W. Simmonds, ed., *Soviet Leaders* (New York: 1967), pp. 248–59; and M. B. Adams, "Lysenko, Trofim Denisovich," in F. L. Holmes, ed., *Dictionary of Scientific Biography* (New York: Charles Scribner's Sons, 1990), vol. 18, suppl. II, pp. 574–78.
5. For the most detailed examinations, see Zh. Medvedev, *The Rise and Fall of T. D. Lysenko* (M. Lerner, Trans.) (New York: Columbia University Press, 1969); D. Joravsky, *The Lysenko Affair* (Cambridge, MA: Harvard University Press, 1970); V. A. Soyfer, *T. D. Lysenko and the Tragedy of Soviet Science* (New Brunswick, NJ: Rutgers University Press, 1994); N. Krementsov, *Stalinist Science* (Princeton: Princeton University Press, 1997); N. Roll-Hunsen, *The Lysenko Effect. The Politics of Science* (Amherst, NY: Humanity Books, 2005); W. deJong-Lambert, *The Cold War Politics of Genetic Research: An Introduction to the Lysenko Affair* (Springer Science & Business Media, 2012).
6. The unit had several technicians, including Lysenko's future wife, Alexandra Baskova, and his lifelong collaborator, Donat Dolgushin.
7. T. D. Lysenko, *Vliianie termicheskogo faktora na prodolzhitel'nost' faz razvitiia rastenii,opyty so zlakami i khlopchatnikom* (Baku: Krasnyi voskhod, 1928). The nearly 200-page book was published as a special issue of the *Proceedings* of the experimental station where Lysenko worked.
8. See R. C. Tucker, *Stalin in Power: The Revolution from Above, 1928–1941* (New York: Norton, 1990); also Sh. Fitzpatrick, ed., *Cultural Revolution in Russia, 1928–1931* (Bloomington: Indiana University Press, 1978).

9. A show-trial of several "bourgeois" engineers accused of "wrecking" the coal mines entrusted to their management. For a detailed, though dated, English language account of the Shakhty trial and its impact, see K. Bailes, *Technology and Society under Lenin and Stalin* (Princeton: Princeton Univ. Press, 1978), 69–140.
10. *Biulleten' Iarovizatsii*, renamed *Iarovizatsiia* (Vernalization) in 1936, and *Agrobiologiia* (Agrobiology) in 1946.
11. See I. I. Prezent, ed., *Khrestomatiia po evoliutsionnomu ucheniiu* (Leningrad: Kubuch, 1934).
12. See T. D. Lysenko & I. I. Prezent, *Selektsiia i teoriia stadiinogo razvitiia rasteniia* (Moscow: Sel'khozgiz, 1935).
13. Th. Dobzhansky, "Vavilov, the martyr of genetics," *Journal of Heredity*, 1947, 38(8):227–32.
14. For published materials of the discussion, see *Sbornik rabot po diskus-sion-nym problemam genetiki i selektsii* (Moscow: VASKhNIL, 1936) and *Spornye voprosy genetiki i selektsii* (Moscow: Sel'khozgiz, 1937).
15. See a general overview of the conference by V. Kolbanovskii, "Obzor soveshchaniia po genetike i selektsii," *Pod znamenem marksizma* (hereafter—PZM), 1939, 11: 96–126.
16. See T. D. Lysenko, "Vystuplenie," *PZM*, 1939, 11: 148–68.
17. See N. I. Vavilov, "Vystuplenie," *PZM*, 1939, 11: 127–41.
18. See M. Mitin, "Za peredovuiu sovetskuiu geneticheskuiu nauku," *Pravda*, 7 December 1939, p. 3. This article, in a slightly revised and expanded form, also appeared in *PZM*, 1939, 10: 147–76.
19. For instance, they published English translations of three main speeches (by Vavilov, Lysenko, and Mitin) at the 1939 discussion. See "Genetics in the Soviet Union: Three Speeches from the 1939 Conference on Genetics and Selection," *Science & Society*, 1940, 4(3): 183–233; see also K. Mather, "Genetics and the Russian Controversy," *Nature*, 1942, 149: 427–30.
20. Th. Dobzhansky to L. C. Dunn, July 4, 1945. Italics are in the original. Dunn Papers in the American Philosophical Society (APS).
21. M. Lerner to L. C. Dunn, June 27, 1945, M. Lerner papers in APS.
22. See T. D. Lysenko, *Heredity and its Variability* (New York: King's Crown Press, 1946); and P. S. Hudson & R. H. Richens, *The New Genetics in the Soviet Union* (Cambridge, Engl.: W. Heffer & Sons, Ltd., 1946).
23. See Kartman, 1945 and L. C. Dunn, "The New Genetics in the Soviet Union," *Science*, 1946, 104: 377–78.
24. See I. Laptev, "Antipatrioticheskie postupki pod flagom 'nauchnoi kritiki'," *Pravda*, September 2, 1947, pp. 2–3; an English translation, see "'The truth' about genetics," *Journal of Heredity*, 1948, 39(1): 18–21.

25. Cited from Trofim D. Lysenko, "Concluding Remarks," in idem, *The Situation in Biological Science* (New York: International Publishers Co. 1949), p. 605.
26. See, for instance, R. C. Cook, "Lysenko's Marxist Genetics: Science or Religion," *Journal of Heredity*, 1949, 40(7): 169–202; J. L. Davies, *Russia Puts the Clock Back* (London: Victor Gollancz Ltd., 1949); and J. Huxley, *Heredity East and West: Lysenko and World Science* (New York, NY: Henry Schuman Inc., 1949).
27. See C. Zirkle, *Death of a Science in Russia* (Philadelphia, PA: University of Pennsylvania Press, 1949); and M. I. Lerner, *Genetics in the USSR: An Obituary* (Vancouver: University of British Columbia, 1950).
28. On the construction of Akademgorodok, see P. Josephson, *New Atlantis revisited: Akademgorodok, the Siberian city of science* (Princeton, N: Princeton University Press, 1997).
29. See, for instance, E. W. Caspari & R. E. Marshak, "The Rise and Fall of Lysenko," *Science*, 1965, 149: 275–78.
30. Ever since his Western colleagues had learned about Vavilov's tragic fate in 1945, they explicitly blamed Lysenko for his arrest and his death. See Vavilov's obituaries by Dobzhansky, Darlington, and others.
31. M. Popovskii, "Tysiacha dnei akademika Vavilova," *Prostor* (Alma-Ata), 1966, nos. 7–8; S. Reznik, *Nikolai Vavilov* (Moscow: Molodaia gvardiia, 1968). The "tamizdat" versions of both books published in the 1980s in the West were much more detailed and explicit on the subject.
32. The manuscript was smuggled out of the USSR and soon appeared in English and French translations, see Zhores Medvedev, *The Rise and Fall of T. D. Lysenko* (New York: Columbia University Press, 1969); and Jaurès Medvedev, *Grandeur et chute de Lyssenko* (Paris: Gallimard, 1971). For Medvedev's being committed to a mental hospital, see Zhores Medvedev and Roy Medvedev, *A Question of Madness: Repression by Psychiatry in the Soviet Union* (New York: W.W. Norton, 1979).
33. See, for instance, A. E. Gaisinovich, *Zarozhdenie genetiki* (Moscow: Nauka, 1967); and idem, *Problemy izmenchivosti i nasledstvennosti v russkoi biologii na rubezhe XIX i XX vekov* (Moscow: Nauka, 1971).
34. See, for instance, N. P. Dubinin, "I. V. Michurin i sovremennaia genetika," *Voprosy filosofii*, 1966, 6: 59–70; and S. I. Alikhanian, *Teoreticheskie osnovy ucheniia Michurina o peredelke rastenii* (Moscow: Nauka, 1966).
35. See, L. Ia. Bliakher, *Problema nasledovaniia priobretennykh priznakov. Istoriia apriornykh i empiricheskikh popytok ee resheniia* (Moscow: Nauka, 1971).
36. See I. T. Frolov, *Genetika i dialektika* (Moscow: Nauka, 1968); and N. P. Dubinin, "Sovremennaia genetika v svete marksistsko-leninskoi

filosofii," in M. E Omel'ianovskii, ed., *Lenin i sovremennoe estestvoznanie* (Moscow: Nauka, 1969), pp. 287–311.

37. See, for instance, memoirs of Lysenko's main opponent in the 1950s, N. P. Dubinin, *Vechnoe dvizhenie* (Moscow: Politizdat, 1973).
38. See E. C. Stakman, "Science and National Authoritarianism," *Proceedings of the American Philosophical Society*, 1950, 94 (2): 114–20; H. J. Muller, "Science under Soviet Totalitarianism," in C. J. Friedrich, ed., *Totalitarianism* (Cambridge, MA.: Harvard University Press, 1954), pp. 233–44.
39. C. Zirkle, *Evolution, Marxian Biology, and the Social Scene* (Philadelphia: University of Pennsylvania Press, 1959), p. 6 and p. 81, respectively. Of course, this reading of the controversy simply repeated Lysenko's own invectives.
40. D. Joravsky, *The Lysenko Affair* (Cambridge, MA: Harvard University Press, 1970), p. VII.
41. L. R. Graham, *Science and Philosophy in the Soviet Union* (New York: Vintage Books, 1974), p. 195.
42. See D. LeCourt, *Lyssenko: histoire réelle d'une "science prolétarienne"* (Paris: Monspero, 1976); an English translation, D. Lecourt, *Proletarian Science? A Case of Lysenko* (London: NBL, 1977); D. Buican, *L'éternel retour de Lyssenko* (Paris: Editions Copernic, 1978); F. Belardelli, "The 'Lysenko Affair' in the Framework of the Relations Between Marxism and the Natural Sciences," *Scientia*, 1977, 112: 33–50; R. Lewontin & R. Levins, "The Problem of Lysenkoism," in H. & S. Rose, eds., *The Radicalization of Science* (London: MacMillan Press, 1976), pp. 32–64; R. Young, "Getting Started on Lysenkoism," *Radical Science Journal*, 1978, 6/7: 80–105; I. Wainright & J. Fredman, "A Re-examination of the Lysenko Controversy," *Proletariat*, 1979, 5(1): 32–44; (2): 6–18; R. Ladous, *Darwin, Marx, Engels, Lyssenko et les autres* (Paris: Institute epistemologie, 1984); J. &. D. Kotek, *L'Affaire Lyssenko* (Bruxelles: Editions Complexe, 1986).
43. See G. Jones, "British Scientists, Lysenko, and the Cold War," *Economy and Society*, 1979, 8(1): 26–58; D. Paul, "A War on Two Fronts: J. B. S. Haldane and the Response to Lysenkoism in Britain," *Journal of the History of Biology* (hereafter—JHB), 1983, 16(1): 1–37; and P. J. Kuznik, *Beyond the Laboratory* (Chicago: University of Chicago Press, 1987).
44. See, for instance, M. Popovskii, *Manipulated Science: The Crisis of Science and Scientists in the Soviet Union Today* (New York: Doubleday, 1979); and J.-P. Regelmann, *Die Geschichte des Lyssenkoismus* (Frankfurt auf Main: Fisher Verlag, 1980).

45. See, for instance, N. Wade, "Gene Splicing: Senate Bill Draws Charges of Lysenkoism," *Science*, 1977, 197: 348–50. Of course, such expansion had already begun during the 1950s anti-Lysenko campaign. See, for instance, L. Mellett, "'Lysenkoism' in Washington," *Journal of Heredity*, 1953, 44: 69.
46. B. D. Davis, "Neo-Lysenkoism, IQ, and the press," *Public Interest*, 1983, 73: 41–59. See also S. J. Gould's response, "Who donned Lysenko's mantle?" *Current Affairs*, 1984, 75: 148–51.
47. See http://www.clubdelhorloge.fr/index.php?option=com_content&view=article&id=33:prix-lyssenko&catid=19:prix-lyssenko&Itemid=129; assessed on October 10, 2014.
48. Of course, the main function of any label is to establish a boundary between different groups of individuals, ideas, practices, actions, public images, etc.—most often, between "us" and "them." It is the *boundary*, not the actual composition (content) of a group (the stuff that boundary encloses), that define the group in any labeling. Thus it should not be surprising that the label "Lysenkoism" could be attached to anyone, including geneticists. In this respect, the polemics between Davis and Gould cited above are as illustrative and illuminating, as is the list of the winners of "Lysenko Prize."
49. M. B. Adams, "Genetics and the Soviet Scientific Community, 1948–1965." Unpublished PhD diss., Harvard University, 1972; idem, "Biology after Stalin: A case study," *Survey: A Journal of East/West Studies*, 1977–8, 23: 53–80; idem, "Science, Ideology, and Structure: The Kol'tsov Institute, 1900–1970," in L. Lubrano & S. Solomon, eds., *The Social Context of Soviet Science* (Boulder, CO: Westview Press, 1980), pp. 173–204.
50. See A. E. Gaissinovich, "The origins of Soviet genetics and the struggle with Lamarckism, 1922–1929," *JHB*, 1980, 13(1):1–51; idem, "Contradictory Appraisal by K. A. Timiriazev of Mendelian Principles and its Subsequent Perception," *History and Philosophy of the Life Sciences*, 1985, 7: 257–86.
51. N. Roll-Hansen, "A New Perspective on Lysenko?" *Annals of Science*, 1985, 42: 261–78.
52. D. R. Weiner, "The Roots of 'Michurinism': Transformist Biology and Acclimatization as Currents in the Russian Life Sciences," *Annals of Science*, 1985, 42: 243–60.
53. J. Sapp, *Beyond the Gene: Cytoplasmic Inheritance and the Struggle for Authority in Genetics* (New York: Oxford University Press, 1987).

54. See "Kruglyi stol. Stranitsy istorii sovetskoi genetiki v literature poslednikh let," *Voprosy istorii estestvoznaniia i tekhniki* (hereafter—*VIET*), 1987, 4: 113–23; 1988, 1: 121–31; 2: 91–112.
55. See, for instance, S. P. Efroimson, "O Lysenko i lysenkovshchine," *VIET*, 1989, 1: 79–93; 2: 132–47; 3: 96–111; V. D. Esakov, ed., "Iz istorii bor'by s lysenkovshchinoi," *Izvestiia TsK KPSS*, 1991, 4: 125–41; 6: 157–73; 7: 109–21; A. A. Liubishchev, *V zashchitu nauki. Stat'i i pis'ma* (Leningrad: Nauka, 1991); and many others.
56. See, for instance, a two-volume series of collected works entitled "Oppressed Science," M. G. Iaroshevskii, ed., *Repressirovannaia nauka*, vol. 1–2, (Leningrad, St. Petersburg: Nauka, 1990, 1994); also *Tragicheskie sud'by: Repressirovannye uchenye Akademii nauk SSSR* (Moscow: Nauka, 1995); *Repressirovannye geologi* (Moscow-St. Petersburg, 1999); and many others.
57. See, for instance, V. A. Soyfer, *Vlast' i nauka. Istoriia razgroma genetiki v SSSR* (Ann Arbor, Mich.: Hermitage, 1989); published in English as *T. D. Lysenko and the Tragedy of Soviet Science* (New Brunswick, N: Rutgers Univ. Press, 1994); a special issue of *Quarterly Review of Biology* on Lysenkoism with recollections by Russian and Polish biologists compiled by US botanist Bentley Glass, "The Grim Heritage of Lysenkoism: Four Personal Accounts. I. Foreword," *Quarterly Review of Biology*, 1990, 65(4): 413–21; a revised version of N. P. Dubinin's memoirs, *Istoriia i tragediia sovetskoi genetiki* (Moscow: Nauka, 1992); a voluminous tome, titled the "Heroes and Villains of Russian science," by S. Shnol', *Geroi i zlodei rossiiskoi nauki* (Moscow: Kron-Press, 1997); V. Birstein, *The Perversion of Knowledge: The True Story of Soviet Science* (Boulder, Col.: Westview Press, 2001); and many others.
58. See a call for developing the new social history of Soviet science in D. A. Aleksandrov and N. L. Krementsov, "Opyt putevoditelia po neizvedannoi zemle: predvaritel'nyi ocherk sotsial'noi istorii sovetskoi nauki (1917–1950-e gody)," *VIET*, 1989, 4: 67–87; and idem, "Sotsio-kul'turnye aspekty razvitiia sovetskoi nauki v 1920–1930-e gg.," ibid., 1990, 1: 166–68.
59. See, for example, a special issue on "New History of Soviet Science" of *Science in Context*, 2002, 15 (2); and a special issue on "Intelligentsia Science: The Russian Century, 1860–1960" of *Osiris*, 2008, 23 (1).
60. See N. L. Krementsov, "Ot sel'skogo khoziastva do…meditsiny," in *Repressirovannaia nauka* (Leningrad: Nauka, 1991). Vol. 1, pp. 91–113; idem, "Ravnenie na VASKhNIL," in *Repressirovannaia nauka* (Leningrad: Nauka, 1994). Vol. 2, pp. 83–96; M. Konashev, "Lysenkoism pod okhranoi spetskhrana," in *Repressirovannaia nauka* (Leningrad:

Nauka, 1994). Vol. 2, pp. 100–101; and N. Krementsov, "Printsip konkurentnogo iskliucheniia," in *Na perelome: sovetskaia biologiia v 20-30-kh godakh* (St. Petersburg: Nestor, 1997), pp. 107–64.

61. See N. Krementsov, *Stalinist Science* (Princeton, NJ: Princeton University Press, 1997); also K. Rossianov, "Editing Nature: Joseph Stalin and the 'New' Soviet Biology," *Isis*, 1993, 84:728–45; A. Kojevnikov, "Rituals of Stalinist Culture at Work: Science and the Games of Intraparty Democracy circa 1948," *Russian Review*, 1998, 57(1): 25–52; and O. Elina, "Planting Seeds for the Revolution: The Rise of Russian Agricultural Science, 1860–1920," *Science in Context*, 2002, 15 (2): 209–37.
62. For instance, it has traditionally been assumed that Lysenko initiated and organized the 1936 and 1939 discussions to discredit genetics and geneticists. In fact, in the 1930s, Lysenko did not need any public discussion of his scientific views: by that time the support of the agricultural bosses had already enabled him to put his ideas into practice, to seize scientific institutions, and to promote his allies to various high-level posts within the agricultural and scientific hierarchy.
63. See, for instance, M. B. Konashev, "Na poprishche klevety protiv sovetskikh biologov (kritika F. G. Dobrzhanskim lysenkoisma)," *Trudy Sankt-Peterburgskogo obshchestva estestvoispytatelei*, 1994, 90 (1): 60–74; N. Krementsov, "A 'Second Front' in Soviet Genetics: the International Dimension of the Lysenko Controversy," *JHB*, 1996, 29(2): 229–50.
64. N. Krementsov, "Lysenkoism in Europe: Export-Import of the Soviet Model," in M. David-Fox and G. Peteri, eds., *Academia in Upheaval: Origins, Transfers, and Transformations of the Communist Academic Regime in Russia and East Central Europe* (New York: Garland Publishing Group, 2000), pp. 179–202.
65. See, for example, E. Hoxtermann, "'Klassenbiologen' und 'Formalgenetiker'. Zur Rezeption Lyssenkos unter den Biologen in der DDR," *Acta Historica Leopoldina*, 2000, 36: 273–300; E. Pollock, *Stalin and the Soviet Science Wars* (Princeton: Princeton University Press, 2006); P. Köhler, "Łysenkizm w botanice polskiej," *Kwartalnik Historii Nauki i Techniki*, 2008, 53(2):83–161; and many others.
66. See F. Cassata, *Le due scienze. Il "caso Lysenko" in Italia* (Turin: BollatiBoringhieri, 2008); W. deJong-Lambert, *The New Biology: Lysenkoism in Poland* (Saarbrucken: VDM Verlag, 2005); idem, "From Eugenics to Lysenkoism: The Evolution of Stanisaw Skowron," *Historical Studies in the Natural Sciences*, 2009, 39(3): 269–99; idem, "The 'Spitzer Affair': Genetics, McCarthyism and the Cold War," in J. Suchoples and K. Turton, eds., *Forgotten by History: New Research on Twentieth Century*

Europe and America (Berlin: LIT Verlag, 2009), pp. 71–85; R. Hagemann, "How Did East German Genetics Avoid Lysenkoism?" *Trends in Genetics*, 2002, 18: 320–24; O. S. Harman, "C. D. Darlington and the British and American Reaction to Lysenko and the Soviet Conception of Science," *JHB*, 2003, 36: 309–52; K. Iida, "Practice and Politics in Japanese Science: Hitoshi Kihara and the Formation of a Genetics Discipline," *JHB*, 2010, 43: 529–70; A. V. Kouprianov, "The 'Soviet Creative Darwinism' (1930s–1950s)," *Istoriko-biologicheskie issledovaniia*, 2011, 3(2): 8–31; N. Krementsov, *International Science Between the World Wars: The Case of Genetics* (London: Routledge, 2005); idem, "Eugenics, Rassenhygiene, and Human Genetics in the late 1930s," in S. G. Solomon, ed., *Doing Medicine Together: Germany and Russia between the Wars* (Toronto: University of Toronto Press, 2006), 369–404; N. Roll-Hunsen, *The Lysenko Effect. The Politics of Science* (Amherst, NY: Humanity Books, 2005); E. Schandevyl, "Soviet Biology, Scientific Ethos and Political Engagement: Belgian University Professors and the Lysenko Case," *Journal of Communist Studies and Transition Politics*, 2003, 2: 93–107; L. Schneider, *Biology and Revolution in Twentieth Century China* (Lanham, MD: Rowman and Littlefield, 2003); A. J. Wolfe, "What Does It Mean to Go Public? The American Response to Lysenkoism, Reconsidered," *Historical Studies in the Natural Sciences*, 2010, 40: 48–78; and many others.

67. For an overview, see W. deJong-Lambert & N. Krementsov, "On Labels and Issues: The Lysenko Controversy and the Cold War," *JHB*, 2012, 45 (3):373–88. Some contributions to this conference appeared in a special issue of the *Journal of the History of Biology*, see W. deJong-Lambert & N. Krementsov, eds.,*The Lysenko Controversy and the Cold War*, *JHB*, 2012, 45(3); others in various issues of the Russian journal *Istoriko-biologicheskie issledovaniia*.
68. See W. deJong-Lambert, "Lysenkoism in Poland," *JHB*, 2012, 45 (3):499–524; and L. Schneider, "Michurinist Biology in the People's Republic of China, 1948–1956," ibid., 525–56.
69. A. Wolfe, "The Cold War Context of the Golden Jubilee, Or, Why We Think of Mendel as the Father of Genetics," ibid., 389–414.
70. See R. Selya, "Defending Scientific Freedom and Democracy: The Genetics Society of America's Response to Lysenko," ibid., 415–42.
71. See M. Gordin, "How Lysenkoism Became Pseudoscience: Dobzhansky to Velikovsky," ibid., 443–68.
72. See F. Cassata, "The Italian Communist Party and the 'Lysenko Affair' (1948–1955)," ibid., 469–98; and W. deJong-Lambert, "Lysenkoism in Poland," ibid., 499–524.

PART I

The Lysenko Affair: The USSR

Lysenko's Predecessors: The Demchinskys and the Bed Cultivation of Cereal Crops

Olga Elina

As is well-known, Trofim Lysenko's rise to the administrative heights of Soviet agricultural science began with the introduction of vernalization—a technique propagated by its creator as a unique innovation that would increase crop yields.[1] Vernalization thus was hailed as a universal "panacea" in the cultivation of cereal crops throughout the huge territory of the USSR breadbasket. One of the most hotly debated issues in this development is the attitude/responsibility of the expert community of agricultural scientists, which gave the green light to the innovation. In other words, why was Lysenko not stopped at the stage of experimental verification, as routinely happened to others who proposed a panacea? This inevitably leads to some additional questions. To what extent does the procedure of verification depend on agricultural practices, academic preferences, public support, and political context? Is an unequivocal answer even possible in the case of agriculture, where different climatic conditions, levels of mechanization, and so on can lead to different experimental results and conclusions? Finally, can a panacea be harmless? To assist in answering these questions, instances when controversial techniques were, in fact, rejected

O. Elina (✉)
Institute for the History of Science and Technology, Russian Academy of Sciences, Moscow, Russia

W. deJong-Lambert, N. Krementsov (eds.),
The Lysenko Controversy as a Global Phenomenon, Volume 1,
DOI 10.1007/978-3-319-39176-2_2

by the expert community before they could be accepted as panaceas for agricultural practice should be examined.

This chapter presents one such instance: the case of the Demchinskys, a father-and-son team who, at the beginning of the twentieth century, proposed a panacea—a new technique of cereal cultivation in beds (ridges). They insisted that it could increase yields drastically in comparison with traditional field cultivation. As in Lysenko's case, the Demchinskys received support from the public; many landowners reported successful results of bed cultivation. Like Lysenko, the Demchinskys managed to secure the patronage of agricultural officials: the minister of agriculture initiated testing of their technique at agricultural experimental stations. Although some of the reports were positive, most of the stations did not approve the proposal and concluded that the technique was not profitable. Thus, bed cultivation never became a scientifically recommended technique in the Russian Empire.

Bed cultivation and vernalization, as individual manifestations of the general case of an agricultural panacea, unavoidably had much in common in terms of their promotion, their assessment by the scientific community, their ability to secure patronage, and so on. Yet despite their numerous common characteristics, the two "panaceas" had quite different fates. An analysis of these differences could help identify the mechanism which, in one case, stopped the wide implementation of a panacea at a stage when it was still "just a proposal," while, in another, the panacea is given a "green light" and widely implemented. I believe that understanding of this mechanism could help answer the key questions of the "Lysenko Affair": Who was Lysenko—a sincere "panacea promoter" or a devious manipulator? What circumstances led to the introduction of vernalization as a mass technique of farming in the USSR?

Miracles in Agriculture

Famines punctuated the entire history of Russia well into the twentieth century, making the "agricultural question" a major issue. By and large, it can be encapsulated into a question concerning the yield: will we have grain this year or not. Given Russia's climatic conditions that could often lead to crop failure, the question of yield, in popular imagination, was associated not with everyday hard work, but with the metaphysical anticipation of a "miracle." The well-known Russian folktale of the turnip is an apt example of the way peasants would perceive a success in obtaining exceptional yield. In the tale, "an old man sows a turnip." Then a "miracle" occurs: "The turnip grew to be enormous."[2] The tale does not show how hard the peasant worked to grow the turnip; the "miracle" happens effortlessly. The tale implies certain intuitive peasant wisdom that was also embodied in many popular sayings and proverbs, such as "Plough before Assumption, reap an extra sheaf" and "Sow rye in ashes, but wheat at the right time."[3]

In the early nineteenth century, science was pressed into service to solve the "agricultural question." It is not surprising that, in the consciousness of Russian peasants, agricultural science was inevitably endowed with magical qualities; the scientist was regarded as a "wizard." One would expect from science an immediate result, a breakthrough in increasing yields. One would expect nothing short of a miracle, a panacea. Scientists themselves encouraged these mythological expectations. A distinguished agronomist and agricultural chemist, Alexei G. Doyarenko (1874–1958), for example, promised that "agricultural science will provide a three- and four-fold increase in yields."[4] Needless to say, among those landowners and farmers who were amateur agronomists, there were a number of "panacea promoters," as they were wittily called by a contemporary journalist,[5] or "harmless cranks," as David Joravsky characterized them.[6] They tried to solve agricultural problems quickly and radically. The evolutionary pace of step-by-step modernization was too slow to satisfy them. What some of the "panacea promoters" were offering was nothing short of a miracle.

The majority of the "panacea promoters" remained unknown "harmless cranks," for they were unable to attract the attention of the public, the scientific community, and the officialdom. Indeed, the attitude of these three groups toward a panacea would determine its subsequent fate. For example, in 1848, scientists from the Moscow Agricultural Society were compelled to spend time on "trials to verify claims that oats regenerate into rye." Their conclusion was negative.[7] Only a few individuals from the ranks of "harmless cranks" managed to have their names inscribed in the annals of science. One of them, albeit in the narrower field of fruit growing, was Ivan Vladimirovich Michurin (1855–1935), a landowner and amateur plant breeder, who created new varieties of apple, pear, cherry, and rowan, and proposed novel techniques of cultivating grapes, apricots, sweet cherries, and other southern plants in the northern climates—which, to many, were nothing but "miracles."[8]

"Miracle-making" in agriculture was an international phenomenon. In the USA, in the early twentieth century, a controversy arose around the figure of the "plant wizard" Luther Burbank (1849–1926), who developed varieties of stoneless plum, a spike-less cactus, and a white blackcurrant, among many other plants, on his farms near Santa Rosa and Sebastopol in California. Historians of science have repeatedly shown that Burbank owed his renown as a "scientist" to the mass culture taking shape at that time, which transformed the poorly educated fruit grower into a symbol of national scientific achievement.[9] But it is worth noting that the eminent plant breeder and geneticist Hugo De Vries, a contemporary of Burbank's, called the stoneless plum "one of the most glorious miracles of the farm at Sebastopol."[10]

In a list of "miracle-makers," Trofim Lysenko (1898–1976) stands out. If one puts aside for the moment the political and ideological contexts of his rise to prominence, Lysenko appears to be just another "panacea promoter." His many innovations, be it vernalization, branched wheat, or sowing potatoes in summer, were all intended to offer radical and simple solutions to the problem of yields. As numerous scholars have demonstrated, Lysenko's ascent owed much to his widespread public fame, to support from respected academics, and to patronage from the state authorities. Yet, Lysenko's rise is, first and foremost, a "Soviet fairy tale": the success story of the "barefoot academician" is closely tied to the visions of a "new intelligentsia," which first began to be articulated in the 1930s.

The Demchinskys: A Story of Two Enthusiasts

Russia has seen many "panacea promoters." In the first decades of the twentieth century, the name Demchinsky was as famous as Lysenko's would be in the 1930s and 1940s. This father-and-son team undertook to solve the "agrarian question" by means of bed cultivation of cereal crops.

Nikolai Aleksandrovich Demchinsky (1851–1914) was born to a noble family in Saransk (Penza province). Nikolai first attended the First Penza Gymnasium and then graduated from the Nezhin Law Lyceum in the Chernigov province (now Ukraine). He served on the construction of the Orenburg railway, and subsequently on the Moscow-Kursk and South-Western lines. By the mid-1870s, he had begun to gain renown as a journalist and writer. From 1874 on, he headed the railway section of a St. Petersburg daily, *Golos* (The Voice). In 1877, he became a newspaper correspondent at the Danube military operations during the Russo-Turkish War. In 1884, he obtained the equivalent of a doctorate in law from the St. Vladimir University in Kiev.[11] At the end of the 1880s, he began to explore "photo-mechanics": he developed an electrolytic etching technique, and in 1890, invented a bimetallic copper-zinc etching plate. From 1900 onwards, he published weather forecasts and other articles on meteorology, historical sketches,[12] political essays,[13] and fiction[14] in various Russian newspapers, including *Novoe Vremya* (New Time), *Novyi Put'* (New Way), and *Russkii Vestnik* (The Russian Herald).

Like many other representatives of the Russian educated nobility, Nikolai Demchinsky was a universalist thinker of a populist bent, devoting himself to the "well-being of the country." One of his more serious pastimes was climatology. Nikolai Aleksandrovich developed a theory of weather prediction using elements of astrological meteorology, based on the analysis of lunar phases.[15] From 1900 to 1903, he used his own funds and raised money to publish a journal, titled *Klimat* (Climate), in four languages: Russian, German, French, and English.[16] His amateurish theories and predictions were criticized by many academic scientists—among them, Russia's foremost chemist Dmitry Ivanovich Mendeleev. Surprisingly, however, Mendeleev also called Demchinsky a "special talent," "God's gifted personality," and stated that Nikolai Aleksandrovich might yet surprise the scientific world with original discoveries.[17]

Eventually, Nikolai Demchinsky moved on from meteorology to agronomy, the Russian nobility's traditional field of interest. He began his exploits with a general theoretical discussion, stating that the country's future depends on the modernization of agriculture.[18] He soon moved from theorizing to experimenting on plots and fields at his estate. He tried a variety of methods of sowing cereals; some of them highly unconventional.

His son, Boris Nikolaevich Demchinsky (1877–1942),[19] had by this time become a qualified agronomist, graduating from the country's leading agricultural school, the Moscow Agricultural Institute (MAI, formerly known as the Petrovskaya Agricultural and Forestry Academy), and he joined his father's scientific pursuits at the family estate.[20] As far as can be judged from his student dossier, his studies at MAI in 1899–1903 were not marked by excessively profound engagement with the subject matter, as they were punctuated by lengthy breaks and endless examination retakes.[21] Yet, there were recommendations that he be completely freed from taking examinations, which emanated from no less a place than the office of the Agriculture Minister.[22] Moreover, there was even a personal letter from the Minister himself—Aleksey Ermolov—inquiring whether Boris would graduate with good grades.[23]

In 1903, Demchinsky Jr. finished his studies at MAI, but did not hurry to find permanent employment. He might be encountered in a members-only chess club in St. Petersburg where, thanks to his "exquisite inscrutability" and brilliant play, he made an indelible impression on the young Sergey Prokofiev, with whom he would become close friends.[24] Or he might be presented to the public as "a young journalist, who had completed a tour of the theatre of the military operations in Manchuria," and gained notoriety thanks to a book about his experiences, titled *Russia in Manchuria*.[25] But even though he did not seek professional employment, he did not neglect agronomy. On the contrary, as soon as he had graduated, Boris joined his father in experimenting with cereal crops at the family estate.

According to Nikolai Demchinsky, the shortage of arable land in Russia was the main obstacle to obtaining high yields in the central and northern provinces in particular, and throughout the country as a whole. A standard scientific approach, in Demchinsky's opinion, could not solve this "neglected problem."[26] Nikolai Aleksandrovich stated that after years of experimenting, he had discovered a method that guaranteed an increase in productivity without increasing the area being sown, which, given Russia's "neglected problem," was a true "miracle."

Demchinsky's method, which he called "the bed cultivation of cereal crops" (BCCC, later also called "ridge cultivation"), included transplanting three-week-old rye and wheat plants into beds and then earthing them up in order to encourage abundant tillering. The advantage of this method, according to Demchinsky, was the radical—tenfold—improvement of yield for a given sown area. Boris joined in his father's project, setting up—in his own words —"27 classical experiments, with controls."[27] Most important, he provided a theoretical basis for the BCCC method. Their joint book on the new method, titled "Securing Yields," was first published in 1908, and in just two years, went through ten printings.[28]

The Demchinskys' "theory" postulated that, if conditions were created for vigorous root growth—or in other words, for the greatest development of the root system—then consequently, the greatest tillering of the cereal crop will occur. In this way, they claimed, it would be possible to achieve colossal yields from a minimal use of sowing material and surface area. As part of their "theory," the Demchinskys divided the entire territory of Russia into two climatic zones: the southern zone they held to be arid, while the northern zone they considered to be afflicted by excess moisture in the soil. In the arid zone, they recommended that the plants be transplanted into beds at a level field, without raised mounds. In those regions affected by excessive moisture, their recommendation was to cultivate cereal crops in beds which had to be banked (earthed) up. The BCCC was deemed particularly suitable for the "moist" zone.[29] The BCCC required special tools: cultivators for transplantation of the young plants and for earthing them up. But the Demchinskys were unable to find people willing to take on the large-scale production of such devices in their home country: it was only possible to use machinery produced abroad, mainly in Germany.

The Demchinskys insisted that the origins of their method lay in "folk wisdom" dating back to medieval China, which they had refined and provided a theoretical basis for. This folk wisdom was contrasted with the "Western scientific culture," which, they alleged, obliterated "the living kernel of peasant practice": "...when a scientific analysis touches on this area, time and again it is compelled to confirm that profound scientific foundations underlay the unconscious, purely mechanical traditions of farming practices."[30] According to the Demchinskys, humanity needs to be fed, and its finest minds should engage with this question, drawing on Western science, but simultaneously being guided by Eastern traditions. Wisdom was to be often found among ordinary peasants who are dealing with soil throughout their lives and not among the people of science. True science was bound to pay heed to "peasant agronomy" with its "unconscious, almost instinctive, but frequently sincere convictions, which have formed over the course of hundreds of years in that backward, or at best poorly-educated, section of agricultural tenants."[31]

Western agronomy, in the Demchinskys' view, was preoccupied solely with saturating the soil with various fertilizers. Under these conditions, there occurs "a dangerous degeneration of the roots," "the cereals use almost exclusively the surface layer of the soil." Plants lose "*the independent ability* to obtain sustenance. This is countered by the BCCC, which guarantees the conditions in which the plants' *independence* can flourish" (emphasis added—O.E.). The development of a powerful root system allows the plant to make great use of the supply of moisture and nutrients in the soil. In this way, the strengthening of the root system achieved by the BCCC leads to "an increase in soil fertility in relation to the transplanted and earthed-up cereals." This facilitates the success in "the struggle of the plant itself against a whole range of negative factors," which in turn allows the plant to develop its autonomy more widely, which guarantees vigorous tillering, the basis of high yields:

> *Tillering energy* that arises within the plant is the main criterion for successful cultivation, and at the same time is the dividing line between the widely adopted approach and the suggested innovation. *Tillering energy* makes it possible to direct the progress of practical farming down that path which, due to objections from mainstream science and the lack of success in individual trials, has remained little travelled [emphasis added—O.E.].[32]

Finally, it is worth describing the Demchinskys' presentation of their method in print. Papers about the BCCC were published not only in the scientific press, which is what a professional would have done (bearing in mind that Boris was a qualified agronomist).[33] Articles on the BCCC first appeared in the newspapers mentioned above—*Novoe Vremya* and *Russkii Vestnik*—as well as in other popular media. Moreover, in a book which came out in 1908, the Demchinskys declared the BCCC to be nothing less than a way of "solving the agrarian problem,"[34] without waiting to hear the response of their colleagues. The theory and practice of the BCCC was undoubtedly influenced by the general worldview of Demchinsky Jr. His mystical outlook, combined with his agricultural training, bore a singular fruit. Boris considered it unforgivable to idly exploit nature, calling this "the parasitism of the biological aristocrat on the environment which feeds him."[35]

The publication of the Demchinskys' 1908 treatise produced quite a stir. Most important, the Ministry of Agriculture gave the task of verifying the method to the experimental stations under its aegis. Why did the authorities—the agricultural ministry in particular—take an interest in the BCCC? Russia was never short of "panacea promoters" making suggestions as to how yields could be radically increased. As the archives show, their suggestions sent to "those in charge" were destined to be shelved without response by the agricultural officials.[36]

But the Demchinskys managed to acquire highly placed patrons. There are indications that they had the personal support of Aleksey Sergeevich Ermolov, the Minister of Agriculture himself. Was it merely his official duties that caused the minister to take an interest in the new method and its authors? Was it, as the anonymous author of a short article on the Demchinskys in the first edition of the Great Soviet Encyclopaedia put it, that "the Imperial authorities would grasp at any straw to save the sinking ship of agriculture."[37] Yet in 1906, this "ship" was far from any state of crisis.[38]

In order to answer the question why the Ministry of Agriculture supported BCCC, it is vital to understand the kind of people who headed this institution. However, strange it may seem, the ministry at that time was headed by agronomist scholars who were quite optimistic about scientific progress and believed in the possibility that farming could be perfected with the aid of science. But those same people also shared the traditional view of agriculture as the primordial activity of Russian peasants, an activity founded on their innate wit, intuition, and experience. Minister Ermolov offers a prime example of this kind of dualistic attitude toward agriculture. He was a graduate of the Forestry Institute, having studied under the famous Russian agronomist and agrochemist Aleksandr Nikolaevich Engel'gardt. Developing his teacher's scientific approach, Ermolov authored an array of scientific works on agricultural chemistry and field husbandry.[39] At the same time, he was renowned as a collector of oral folklore related to agricultural matters. The following statement by him is strikingly reminiscent of the Demchinskys' pronouncements on "peasant agronomy":

> In this field [*folk/peasant agronomy—O.E.*], there is a great deal for which science has yet to find a satisfactory explanation, and yet which cannot be refuted *a priori* ... Many instances are known when the power of folk observation has anticipated that which was only later ascertained by science, but which prior to this had not only been ignored by science, but actively repudiated, being considered a traditional, but wrong opinion. From this point of view, then, folk wisdom in many instances only expands the field for future scientific observations, sometimes opening up new paths and horizons ... This is precisely the reason why I ascribe to it great significance.[40]

Ermolov edited and published the material he had collected over the years in a multivolume work, titled "Folk Agricultural Wisdom in Sayings, Proverbs and Omens," which he supplemented with a detailed scholarly analysis of the sources. In the final volume, published in 1905, Ermolov also included a favorable discussion of "the very recent theory proposed by N.A. Demchinsky of predicting the weather on the basis of lunar phases."[41] So it is no surprise that Ermolov took an interest in Demchinsky's new proposal, this time in the sphere of agronomy. The Demchinskys' theory—rooted in "folk wisdom," based on experiments, and supported by a scientific theory—hit the bull's eye. It seems likely that Ermolov himself, or through his assistants, was able to put in place the verification process for the BCCC. But since he retired at the end of 1906, it is difficult to explain the enduring benevolence of the authorities toward the Demchinskys in terms of Ermolov alone.

It turns out that the Demchinskys were also actively promoted by Sergey Yul'evich Witte (also Vitte), the Minister of Finance (1892–1903) and later—the Head of the Council of Ministers (1903–1906). The relationship of complete trust developed between Witte and the Demchinsky. Demchinsky Sr. regularly reported to the prime minister in a private capacity on the state of affairs in the frontline forces during the Russo-Japanese War.[42] In his "Memoires" Witte recalled Demchinsky's letter which described his "titanic efforts" to develop and propagate the BCCC method during his three-year travels around Russia.[43] A personal friendship developed between Witte and Demchinsky Jr., too. The young journalist was entrusted with promoting Witte's book *The Origin of the Russo-Japanese War* intended to whitewash Witte's role in the war.[44]

But even this patronage was not the most powerful. Demchinsky Sr. also enjoyed support from the highest possible position: the Tsar. It has not yet proved possible to reconstruct in detail the history of how Demchinsky was able to enter such elevated circles. Apparently, it was connected with his weather forecasts and his magazine, *Climate.* Several documents seem to suggest that he provided the royal family with weather forecasts and horoscopes. A biographical article on Demchinsky Sr. in the dictionary "Activists of the Revolutionary Movement in Russia" states: "Nikolai Aleksandrovich Demchinsky was close to the court of Nicholas II, who was interested in his (weather) predictions and financed the publication of *Climate.*"[45] Moreover, there is reliable archival material attesting to the fact that Nicholas II granted Demchinsky Sr. a personal audience and permission to address the Tsar "at any moment and about any matter." According to historian V. M. Bokova, Demchinsky convinced the Tsar of the necessity of certain modernization projects and even drafted some proposals, including, most astonishingly, a project to encourage peasants to build a new type of house instead of the traditional *izba.*[46]

Regardless of how well-disposed the upper echelons of Russian society were toward the Demchinskys, the success of the new method depended on the opinion of experts. And since the method was widely trumpeted in the media and thus was available for use on private farms, its success also depended on its reception by the public.

The public reacted instantly and vigorously to the BCCC, thanks to the Demchinskys actively advertising their theory in the press, in public debates, and at demonstration sites. Nikolai Aleksandrovich reported that he had over 1000 supporters who had successfully conducted trials on their own land and "have been using the BCCC method with outstanding results." The book, *Securing Yields*, which presented the theory behind the BCCC, became a bestseller, running into ten printings between 1908 and 1909.

The method's most faithful supporters happened to be retired officers. Demchinsky Sr. wrote:

> To the honor of military men, I have to say that I have had literally thousands of exchanges on the organization of trials. Hopes are also laid on rural priests, teachers, and, of course, the peasantry. The latter, incidentally, are the hardest of all to persuade that they should conduct independent trials, and so a landowner should come and set up a trial on his own land in full view of his neighbors the peasants.[47]

The scientific community approached the theory with less enthusiasm. While the Demchinskys did have some scholarly defenders, the majority met the method with their weapons drawn. Each public appearance made by the Demchinskys turned into a lively exchange of views. At one such debate, held at a meeting of the Russian Agronomists' Mutual Aid Society, the famous Russian agronomist and plant physiologist Dmitriy Nikolaevich Pryanishnikov severely criticized the Demchinskys' theory and concluded: "Here, the authors of this novelty have paid for their fabrications with a shameful defeat, but who will pay for those huge losses which this undertaking has inflicted on farming?"[48] Pryanishnikov most likely was thinking of those farmers who had used the BCCC method at their own peril. But the BCCC also underwent trials at the Imperial network of agricultural experimental stations for a number of years, which came at an undeniable cost to the country, in terms of scholarly time and taxpayers' money.

The BCCC was tested at approximately 40 experimental stations under the aegis of the Ministry of Agriculture. A few stations did report neutral or positive results of the trials. For example, the Novozybkov Agricultural Experimental Field, in Chernigov province, reported that "the method has proved itself and may be employed on large farms which are sufficiently mechanized."[49] Here, two different sets of factors could have been at play. The first is the specific conditions in Chernigov province that contained sufficiently large private agricultural holdings, which could draw on relatively cheap labor and which were well-equipped (by Russian standards) with necessary machinery.[50] It seems also possible that agronomists at the Novozybkov Agricultural Experiment Field were either personally acquainted with, or had indirect links to the Demchinskys. In fact, an estate that belonged to the brother of Demchinsky Sr. was situated in the Sosnitsky district of Chernigov province. And it was at none other than the Novozybkov agricultural station that Demchinsky Jr. began his official career in agronomy. So it is possible that the neutral, if not downright positive, assessment of the BCCC at this station was motivated by certain personal considerations of its staff.

According to the data from the majority of stations, transplanting and earthing up cereals did not produce positive results. The method was found to be excessively labor-intensive and, therefore, not profitable: the increase in yield did not cover the colossal costs incurred in transplanting and earthing up the crop. As a result of careful research conducted at the Shatilov Agricultural Experiment Station in 1908, the method was "totally condemned." Petr Ivanovich Lisitsyn, the official in charge of the experiments, wrote:

> In conclusion, I cannot refrain from addressing the moral aspect of the work. It seems to me (and, most likely, everyone who has taken a part in conducting trials of the Demchinsky method will agree) that the transplantation [of young plants from the original sowing locales to the mound—O.E.] would serve as an ideal form of hard labor. Despite all the energy with which Demchinsky went about successfully gaining the interest of several government departments in his mound cultivation, he has made a grievous error in failing to interest the prisons department. There, his cultivation method would have a bright future. This transplantation could successfully poison the prison term of even the most hardened criminal.[51]

Of course, the authors of the method were not convinced. Boris Demchinsky, for instance, declared, "The vehemence introduced into this question by all the preceding polemics has led to a situation where the trials have been carried out without sufficient objectivity. These were not trials of the sowing method so much as burials of it."[52] Despite some positive assessments, the overall negative conclusion of the experts proved quite convincing for the Ministry of Agriculture. There was no suggestion whatsoever that this method, which had failed to gain acceptance, should be "deployed everywhere."

However, not everything to do with the BCCC was as clear-cut as Pryanishnikov and Lisitsyn would have liked. In fact, the Demchinskys' ideas aroused great interest in several European countries, both in scholarly circles and among farm owners. There, the history of BCCC had lasting practical impact. The Demchinskys' 1908 book almost instantly appeared in German, and somewhat later in English and French translations.[53] Numerous articles about their method were written across Europe, notably in the UK.[54] According to the Dutch scholar Dominic Glover, in 1911, A.E. Parr, Deputy Director of Agriculture for the United Provinces (UK), published an article discussing his own experiments with the Demchinsky method that had "created so much interest in the agricultural world of Continental Europe during these last three years."[55] The method had its widest following in Germany, where a special cultivator for earthing up the cereals is known to have been designed and manufactured. Taken as a whole, from the point of view of Western experts, the Demchinsky method did indeed require too much expenditure to be considered effective. Yet, it is far from deserving of the label "fiction" and is perfectly well-suited to cultivation, albeit under specific conditions. For example, it works well for the cultivation of rice in countries with cheap labor.[56] Provided that a holding is large and the farming process is highly mechanized, the BCCC could indeed produce good results.

Clearly, this in no way applies to that country where the Demchinskys had dreamed they would bring happiness first. Nonetheless, bed cultivation was practiced on an array of farms in Russia, too. It was regularly presented at the annual agricultural exhibitions in the early 1910s. But the death of Nikolai Demchinsky in 1914 and the beginning of World War I put the question of its implementation on the back burner.

The BCCC: A Soviet History

The Bolshevik coup d'état of October 1917 led to vast changes in Russia, but not so much in the science of agronomy: the Imperial network of agricultural experimental institutions ended up under the aegis of the People's Commissariat of Agriculture (Narkomzem), which was created in place of the Ministry of Agriculture.

As for Boris Nikolaevich Demchinsky, nothing was heard of him for a long time. Keeping a low profile, he worked quietly as an agronomist at the Novozybkov station in the remote Chernigov province. Sometime between 1921 and 1923, he moved to Petrograd, where at the end of 1923, he was arrested and imprisoned. The details of his case—where exactly and why he was arrested—are unknown. What is known, though, is that a letter was sent from "cell 246" of the infamous Kresty Prison by a special messenger to Narkomzem. It bore the stamp of the deputy head of the Petrograd Department of the secret police (OGPU) and was written by Boris Nikolaevich Demchinsky.[57] The letter contained a request that Narkomzem "order another trial program for the bed cultivation of cereal crops method."[58] This time, the trials were to be held at Soviet experimental stations, mainly in the north of the arable region of the USSR, Central Russia. Demchinsky emphasized that he did not "plan to extract any profit, either directly or indirectly" and that he was asking nothing for himself, he was "only thinking about the idea … that could profit [the country]."[59] It is noteworthy that a significant portion of Demchinsky's letter was devoted to references to the attention the method had received in Britain, France, and Italy, and to its allegedly successful application in Germany.[60]

Demchinsky also managed to send a parcel to Narkomzem (presumably via his friends) with two of his books: the 1908 *Securing Yields* and the recently published *Retribution for Cultivation: The Movement to a Worldwide Famine*, a new reiteration of his ideas about the two paths that agricultural development might take. In this book, he was even more categorical about the advantages of his panacea:

> The lack of attention to the method [*BCCC—O.E.*] I am propounding would lead to "retribution for cultivation practices": the cultivation of plants has imperceptibly turned into a cult of plants … Plants have been losing their *autonomy*: the overfeeding of plants through saturating the soil with various fertilizers has been taken to its logical conclusion … What was to blame for this was nothing less than the entirety of modern science: not only technology, but also agronomy, and the latter more than anything is filled with a hankering after the most speedy transformation of autonomous plants into parasites … Agronomy, trampling on biological laws, will drive matters all the way to the extinction of agricultural plants and, consequently, to a worldwide famine.[61]

In his letter to Narkomzem, Boris Nikolaevich recapped the main points of his "retribution for cultivation practices" concept.

The letter did not make its way to the very top, the People's Commissar of Agriculture. Demchinsky's proposal was examined by the scientific subsection of Narkomzem's Agricultural Experimental Section. The head of this subsection, agronomist Mikhail Arnol'd, wrote a report that, in part, stated:

> The overall opinion of the testers had not come down in favor of this method of cultivation. The Experimental Section believes that, no matter what the technical effect of employing the bed cultivation of cereals, it cannot, under the Russian conditions, have practical significance, as a result of

> its likely unprofitability. Nonetheless, it appears only desirable to subject this question once again to careful verification. Therefore, ... the Section considers it possible ... to make enquiries of every agricultural experimental station in the northern regions of the RSFSR as to their willingness to conduct trials on their fields into earthing up cereals according to the Demchinskys method.[62]

The notation—"Agreed"—was added by M.F. Shefler, the head of the Farming Department and member of the Narkomzem Collegium. This meant that the program of trials would begin anew. Of course, the first step of the program was to release Demchinsky from prison and close the case against him. Not every station—in fact, far from it—expressed "willingness" to take part in the trials of the method, it being a voluntary matter. Several, for example, the Shatilov station replied with some indignation that they had already provided a negative assessment of the method. Others agreed to hold trials once again, even though "in 1909, trials had already been conducted into mound cultivation according to the Demchinsky method."[63]

The results obtained from the testing stations once again failed to confirm the profitability of the BCCC. The conclusions were very similar to those that would be made a decade later concerning Lysenko's vernalization of cereals: the increase in yield, if any were actually obtained, would require excessive levels of labor expenditure. It is not necessary to discuss the way in which the Demchinskys' earlier critics—Pryanishnikov, Lisitsyn, Doyarenko, and many others—used all of their influence to prevent any widespread introduction of the BCCC. Narkomzem, just like the Imperial ministry, was inclined to accept the opinion of the scientific community, and so, the second attempt to spread the BCCC again came to naught, thanks to expert opinion.

But, as if in deference to the narrative demands of a fairy tale, the Demchinskys and the BCCC would make a third appearance in the early 1930s. This was linked to the launch of a large-scale manufacture of mechanical cultivators, designed to plant potatoes and other root vegetables. There was some speculation in the press as to whether these cultivators might be used to transplant cereals according to the Demchinskys method. Encouraged by such attention, Boris Demchinsky came up with a new version of the BCCC and published a new book on *Developments of the Agricultural Industry.*[64] Entirely in keeping with the spirit of the 1930s, Demchinsky suggested that the country of victorious socialism should shun the "dead-end of world science" and "go down a different road" in order to "save humanity":

> To prevent the further degradation of plant crops and agricultural animals is to respond to the interests of all mankind. This is a question which is truly worldwide in scope. Who, then, can put their hands to its resolution? Of course, only the USSR can. The USSR, which looks beyond the present day and which does not limit its outlook to the boundary markers that bear witness to private interests, but sees before it the whole of humanity.[65]

This time, the method encountered a serious critic from that camp of scholars who enjoyed the particular support of the Soviet authorities: Vasily Robertovich Vil'yams (also Viliams). Viliams had offered his own "panacea" in the sphere of cereal crops cultivation: grassland farming system. Soon thereafter, Trofim Lysenko introduced his method of vernalization. Boris Demchinsky abandoned the BCCC, moved to a remote experimental station in the far north, and undertook research on growing plants in permafrost.[66]

From the Demchinskys to Lysenko

The BCCC and vernalization had much in common in terms of promotion by their respective authors, their assessment by the scientific community, and their support by agricultural officials. Why, then, did the two techniques have such different fates? As materials presented in this paper indicate, both methods were proposed as panaceas, i.e., their role in potentially boosting yield was greatly exaggerated. Indeed, the experts used the very term "panacea" in discussing the two methods (of course, in quotes—that is, with a negative connotation). In the works of Lysenko himself, I have not found the use of this term. However, Lysenko's contemporaries defined the attempts to introduce vernalization as a mass method in terms of "panacea." Some of them warned against these attempts precisely because the method *was not an actual panacea*. This is exactly how the agronomist and plant breeder Andrey Sapegin characterized it: "the method proposed by Lysenko, is not universal, it is not a panacea."[67]

Neither the Demchinskys nor Lysenko created their methods from scratch. The Demchinskys stressed "oriental peasant roots" of their panacea, pointing to the origins of the BCCC in ancient Far-Eastern agricultural practices. With regard to vernalization, the phenomenon was known (and even occasionally used in farming) as far back as the nineteenth century. Lysenko was the first only in the sense of introducing the new term *yarovizatsiya*. In the USSR, the influence of low temperatures on plant development had been successfully studied by, *inter alia*, G.S. Zaitsev, N.A. Maksimov, and A.I. Poyarkov.[68] Despite a common belief that Lysenko was upholding his personal authority and avoided mentioning his predecessors (whose works he either ignored or cited incorrectly), in his early works, Lysenko mostly emphasized the fact that he had managed to step much further, developing, on the one hand, a theory of the stage development of plants to explain vernalization and, on the other, precise recommendations of how to use vernalization on different crops.[69]

The two methods had certain grounding in concurrent agricultural research and thus *cannot* be justifiably labeled "pseudoscientific." Both bed cultivation and vernalization could actually work, that is, provide an increase in yield, although their profitability depends on a number of specific conditions, including the nature of the holding (its economic status and level of mechanization), and a range of botanical and geographical factors. When these methods came to be applied in Russia, in those specific regions and with those specific crops, the profit margin was not high enough. "It was not," as the saying goes, "worth the candle."

Despite all the differences in their background and circumstances, the progenitors themselves had many characteristics in common. In both cases, there existed a father–son working team. Just as in the Demchinskys' case, Lysenko's vernalization was presented as an approach tested in practice by a father, who was a self-taught amateur, and given a theoretical basis by a son, who was a professional agronomist. In both cases, the progenitors managed to secure support from on high, as well as to obtain public recognition; the press was of great assistance with the latter. And the most important similarity: the authors of both methods were absolutely intolerant of criticism, considering opponents of their method to be their personal enemies. Fanatical about their own particular brand of science, they were prepared to go to the bitter end in insisting that their panacea be introduced everywhere.

It is quite possible then that both the Demchinskys and the Lysenkos were genuinely convinced of the value of their methods, believing that they had made discoveries that were capable of overturning existing agricultural practices. The problem consisted in the somewhat limited scientific groundings of the junior bearers of these prominent surnames, in spite of their diplomas. It may indeed have been the case that Boris Demchinsky received a sound education in agronomy and the natural sciences, but precisely at the time when he was elaborating the theory behind the BCCC, far from immersing himself in science, he had, in fact, left it behind, engaging instead in journalism and even mysticism. In the case of Trofim Lysenko, the picture was different: it is known that Lysenko delved deeply into the topics that interested him, but he did not have a broad basis to his general education. Over time, as it is clear from the Demchinskys' example, there developed not only fanaticism and stubborn reluctance to heed critical opinions but also obvious exaggerations and distortions of data, when the desired results took precedence over those actually obtained (cf. B.N. Demchinsky's assertion that the method was widely used abroad).

However, the differences between the two methods seem much more important for understanding the success of Lysenko's vernalization in comparison with the failure of the Demchinskys' BCCC. The press greatly contributed to the popularity of both methods, but in fact, it did more for Lysenko, who had nothing in common with the media, than for the Demchinskys—well-known journalists of their time. Despite the fact that the Demchinskys widely published their articles on the BCCC in a number of popular magazines, using their affiliation with the world of journalism, in order to grab the attention of the whole country they had to use other formats of communication. Thus, to promote the BCCC in the provinces, Nikolay Demchinsky had to deal with endless lecture tours and demonstration courses. He complained to his patron and friend, the former Prime Minister Witte, that in the heyday of the BCCC he was unable to drop by his estate, constantly shuttling across the country.[70] Lysenko, in this sense, was much luckier: the press launched the promotion of vernalization method on its own initiative by all available means, beginning with the well-known 1927 article in *Pravda* and culminating in a 1931 article in *Izvestiia* that was apparently inspired by Narkomzem officials.[71] Publications in these newspapers (both with a countrywide circulation) were crucial in generating publicity for Lysenko and his vernalization. The 1931 *Izvestiia* article is quite revealing in this respect:

> We must specifically raise the issue of the works of Comrade Lysenko. Comrade Lysenko himself is still unaware of the importance of his results and their HUGE VALUE … Our task is … during the spring of 1933 to apply the method of Comrade Lysenko on a massive scale, on a territory of at least hundreds of thousands of hectares. Only with such a scale will the matter be put on a truly scientific path, on a really revolutionary mode.[72]

Although both the Demchinskys and Lysenko managed to secure support from on high, the nature of this support in the two cases differed. The Demchinskys secured the "highest patronage" even before they started to promote the BCCC. In the early 1930s, Lysenko did not yet enter the circle of scholars who were supported by Joseph Stalin personally, but he attracted the attention of the officials from Narkomzem. Stalin's patronage of Lysenko began much later than the transformation of vernalization into a "panacea" had been launched. Therefore, we can only mention the external manifestations of the "highest patronage" in the two cases. For example, there were some amusing similarities: Nicholas II personally edited the texts of Nikolai Demchinsky, while Joseph Stalin did the same for Trofim Lysenko.[73] Nevertheless, support from the Tsar meant exclusively private advantage: it gave money and other privileges of proximity to power (e.g., the ability to exhibit Demchihnsky's works at the fairly private Tsarskoselskaya exhibition of 1911), but could not affect the "scientific weight" of the client, nor protect him from scientific criticism. As for Stalin's patronage, at the zenith of his interest in Lysenko, the latter was exempt from any criticism.

The next—and the most important—difference in the social history of the two panaceas is the attitude of the professional community toward the innovation. The professional community was united in its negative attitude toward the Demchinskys' BCCC. The experts operated as they were supposed to, time and again, in both the pre-Revolutionary period and in the Soviet era. While doing necessary experiments, they were gathering records and data, and based on these results, they directed to the Ministry of Agriculture their negative conclusions about the method's practical effectiveness (notwithstanding the existence of some doubts expressed in several reports). Both academic plant scientists (for instance, plant physiologist Dmitriy Pryanishnikov) and agricultural practitioners (for example, agronomist and plant-breeder Petr Lisitsin) assessed the method as unpromising.

In the case of Lysenko, the situation with the expert community was much more complicated. The most astonishing fact is that Lysenko gained approval from the academic community, particularly from the geneticist and plant breeder Nikolai Vavilov. Of course, it is important to clarify the context in which vernalization was accepted by this group of scholars. Vavilov saw in vernalization an opportunity to make much wider use of the resources contained in his famous seed collection. Using vernalization, it would be possible to combine the flowering periods of plants from different regions of the world. In this way, Vavilov thought that vernalization would facilitate more active research with the seed collection.[74] This was precisely the reason why Vavilov and many other scholars working on plant breeding in the Soviet Union and abroad supported the method.[75] In 1931, at the direction of the Lenin All-Union Academy of Agricultural Sciences (VASKhNIL), a range of institutions, including Vavilov's All-Union Institute of Plant Industry (VIR) and the Institute of Plant Protection, made their collections of wheat and other cereals available to Lysenko for his research into vernalization.[76]

When it comes to the use of vernalization as a means of boosting crop yield through mass adoption of this approach in agricultural practice, the attitude toward vernalization from among the agronomists-practitioners was much more ambiguous than that of academic researchers.[77] The practitioners' ambiguous assessment of vernalization was primarily a result of the lack of the method's testing at agricultural experimental stations: the main trials were held at ordinary Soviet farms (*sovkhozy*) and collective farms (*kolkhozy*), where vernalized seeds had been sown and which thus had provided the balk of the data.[78] The first statistic is that, out of 59 farms participating in the trials, ten observed a negative effect (a decline in yield); four noticed no effect; the remaining 35 did observe an increase in yield to a greater or lesser degree, from 50 to 200 kg/ha. With regard to this increase specifically, Lysenko himself only claimed an average yield increase of a little over 100 kg (1 cwt)/ha.[79] It is not easy to believe that manipulation of the data, to which many of Lysenko's critics refer, could truly have happened on such a scale, as to influence the overall positive result. But the main question, which the experts had to answer, was whether this substantial—at first glance—increase in yield was sufficient for the method to be widely adopted, notwithstanding its many deficiencies.

Even such a consistent critic of both the Demchinskys and Lysenko as Petr Ivanovich Lisitsyn did not criticize vernalization during the early period of its implementation in 1929. The method works, he stated, although it is not new. Its main difficulty lies in the inevitable damage to the germinating seed as it is being sown. This can be done harmlessly on a small farm, where care can be guaranteed "when sowing by hand, broadcasting"; on large areas, "a sowing machine is too crude. A specially-designed machine is necessary." It comes as no surprise that the conclusion Lisitsyn reached sounded optimistic: "I am in favor of the practical implementation of Lysenko's idea."[80] However, he instantly went on to observe that long and painstaking work would be needed, which would include perfecting the design of the special machinery. It should be borne in mind that this was exactly the period when hopes for modernization in Soviet Russia's agriculture were bound up with a breakthrough that was primarily technological; foreign machinery was beginning to arrive at Soviet fields, while Soviet factories for the production of agricultural machinery were being built in cooperation with foreign specialists.[81]

Further trials, including several at experimental stations, followed while vernalization was already being widely practiced around the country. It is possible that, in some regions, the trials were not conducted with much care. In eastern Ukraine, "vernalization did indeed work," a result that was observed more than once. But many stations reported negative results.[82] It is likely that the notes from the logbooks of the agricultural experimental stations, fields, and farms where the efficacy of vernalization was being tested, could provide a different picture, but they are yet to be analyzed in detail.

One could suggest that in many cases, the experts—some of them conducted the tests themselves—were preoccupied with the anticipation of success, in the sense of that "straw" which is grasped at during a period of crisis. Indeed, the situation in agriculture in the Soviet Union in the beginning of the 1930s was critical. The Volga region and Ukraine suffered severe famine induced by the forceful collectivization of the peasantry. As Joseph Stalin himself noted, "the years of the highest filament of the reconstruction in agriculture –1931 and 1932–were the years of the greatest decrease in crop production."[83] The average yield during the period of 1928–1932 was only 7.5 cwt per hectare, which was much lower than what had been projected in the Communist Party's plans. The pressure being brought to bear on the community of agricultural scientists by Narkomzem, which was demanding immediate acceleration of measures to improve the farming practice, should not be forgotten.[84] Shortage of time for experimental verification and the necessity to judge with limited data—is another explanation for the ambiguous conclusions of the expert community. Furthermore, the agronomic community in Russia, being traditionally populist-oriented, deeply empathized with starving villages, and sought to increase yields by all means. The official slogan of the time (put forward not by "panacea promoters" like Lysenko, but by respected professors of the Timiryazev Agricultural Academy) proclaimed: "Soviet agricultural science knows the way to doubling and tripling of yields."[85] Under such complicated conditions, the social responsibility of agronomists-experts could have affected their assessment of a "panacea"—since underestimation of a "panacea" meant the continuation of the tragedy of famine. In addition, the method seemed to be of little expense: all that was needed was just the labor of peasant women (to stir the grain) who were usually not so busy during the winter/early spring period. Vernalization became, in the direct sense of the word, the great hope of immediate and cheap increase in agricultural productivity, the anticipated "Soviet panacea."

Thus, the experts did not shut down the method of vernalization—as they had the BCCC method. Their ambiguous assessment might have been motivated by the expectation of the rapid mechanization of agriculture, or by acute social responsibility for the poor harvests during the first years of collectivization. One should not underestimate the request for a panacea from power structures, who sought to resolve the agricultural crisis caused by the collectivization. One way or another, the expert community did not give (and, in the complex conditions of Soviet Russia at the beginning of the 1930s, it scarcely could have given) an unequivocal assessment of Lysenko's panacea. And this was enough for the mass publicity of the method by the press and the subsequent "rise" of Lysenko's scientific credentials: he was transferred to a specially created Vernalisation Department in the All-Union Institute of Genetics and Plant Breeding in Odessa, while the publication of his newly established journal *Vernalization* helped further promote his ideas. Furthermore, a mass movement around the panacea was founded. Soon thereafter, a theoretical explanation of the method was presented by Lysenko as part of a *new science of agrobiology*.[86] As Elena Sheehan observes, it marked an important stage in Lysenko's evolution from the progenitor of a simple method to becoming the author of a full-scale biological theory.[87]

The answer to the question of whether Lysenko, the Demchinskys, and many others actually were convinced "panacea promoters" or cynical manipulators trying to foist a fiction onto the public is far from unequivocal. As I have tried to show in this chapter, the truth, as it so often happens, lies somewhere in the middle. Champions of a panacea often start with good intentions and motives: "to feed the country." But they inevitably move to the stage of manipulation, evolving from "harmless cranks" into dangerous propagandists for whom all means are acceptable.

And, finally, the key question: who was responsible? The success or failure of a panacea lies in the hands of the expert community, which, regardless of the obvious complications of a multifaceted agronomic experiment, the pressure of circumstances and the powers-that-be, is the only body capable of making sense of the proposed innovation and deciding its fate. Thus, formally, *the responsibility for launching of Lysenko's panacea lies with the experts.* However, specifically in the case of vernalization, it is hardly possible to blame them: to ban such a simple and cheap method and thus take away any hopes from the starving villages of the early 1930s was not the same as to give an adequate assessment for "tedious, labor-absorbing" transplantations advocated by the Demchinskys during the late 1900s.[88] It sounds very pessimistic, but my conclusion is that in the early 1930s, Soviet Russia was doomed to the vernalization panacea.

***

P. S. While working on this chapter, I have learned with great surprise that a modification of the BCCC (the so-called "ridge tilling method," based on similar technique of cereals cultivation in beds/ridges) is used in modern-day practice of wheat and other cereals cultivation. However, it is used not in excessively humid regions, as was recommended by the Demchinskys, but to retain moisture and maintain a loose soil structure in irrigation regions with very complex climate conditions (in southern Kazakhstan and Kyrgyzstan, in some parts of Mexico, and the USA). I cannot discuss the profitability of the method and other practical issues, not being an expert in modern agricultural production. However, a number of agricultural scientists in Mexico, the USA, Kazakhstan, and other countries confirm the effectiveness of the method for local conditions in numerous publications.[89] An important component of the success of the method is agricultural machinery specially designed for bed/ridge cultivation. Does this all mean that the Demchinskys were right, and their critics were mistaken in the assessment of the method? Of course not. What is now called "ridge tilling" is a much simpler method and is not as labor-consuming as the BCCC. And yet, as the Demchinskys predicted, the availability of specific machinery led to the revival of the discussion concerning the use of "marginal" methods in agriculture. As for vernalization, nowadays, its applications are limited to specific scientific purposes. Vernalization—or "the activation of a plant's diapausa with cold," as this method is called in today's ecological physiology—is used in breeding practices, allowing subsequent generations of plants to be obtained in selection chambers, phytotrons, and hothouses, and for the flowering period of plants to be combined. Exactly this application of the method was the key element in the approval of vernalization by academic plant breeders, contemporaries of Lysenko, including Nikolai Vavilov.

Notes

1. Vernalization is a method of cooling seeds in a controlled manner prior to sowing them, which allows the plant's period of vegetation to be brought to an end and, in certain instances, enables spring crops to ripen in colder climates.
2. M.K. Azadovsky. N.P. Andreev, Yu.M. Sokolov, eds. 1936. *Narodnye Russkie Skazki A.N. Afanas'eva*. (Moscow: Academia), vol. I, p. 145.
3. Dal', Vladimir I. 1994. *Poslovitsy i Pogovorki Russkogo Naroda*. 3 vols. (Moscow:). vol. 3, pp. 513, 544.
4. *Rossiiskii Gosudarstvennyi Arkhiv Ekonomiki* (hereafter—RGAE). F. 478. Op. 5. D. 2397. L. 170.
5. The Russian term for "panacea promoters" was "*radeteli o panatsee*," according to the anonymous author. See *Tsentral'nyi Gosudarstvennyi Arkhiv goroda Moskvy* (hereafter—TsGA Moskvy). F. 184. Op. 4. D. 170. L. 4–5.
6. Joravsky, David. 1986. *The Lysenko Affair* (Chicago; London: University of Chicago Press), p. 39.
7. TsIAM. F. 419. Op. 1. D. 837. L. 2.
8. See Michurin, Ivan V. 1932. *Itogi Poluvekovykh Rabot*. (Moscow: Gosizdat Sel'skokhoz. i Kolkhoznoi Literatury). 2 vols.
9. Pandora, K. 2001. "Knowledge Held in Common: Tales of Luther Burbank and Science in the American Vernacular," *Isis*. 92 (3): 484–516.
10. De Vries, Hugo M. 1907. *Plant-breeding; Comments on the Experiments of Nilsson and Burbank*. p. 161.
11. According to other sources, he graduated from the Novorossiysk University in Odessa, see Brokhaus i Efron, 1993. "N.A. Demchinsky." *Entsiklopedicheskii Slovar'. Biografii*, 4: 617–618.
12. Demchinsky, Nikolai A. 1905a. *Chto Takoe Obshchina?* (St.Petersburg); idem., 1905b. *O Narodnom Predstavitel'stve*. (St.Petersburg: Knigoizdatel'stvo "Druzei Svabody i Poryadka").
13. Demchinsky, Nikolai A. 1905c. *Otkrytoe Pis'mo k Izbiratelyam v Gosudarstvennuyu Dumy* (St.Petersburg); idem, 1905d. *Chego Khotyat Lyudi, Kotorye Khodyat s Krasnym Flagom* (St.Petersburg).
14. Brokhaus i Efron, 1993. "N.A. Demchinsky." *Entsiklopedicheskii Slovar'. Biografii* 4: 617–618.
15. See Demchinsky, Nikolai A. 1900. *Vozmojnost' Tochnogo Predskazaniya Pogody* (St. Petersburg); idem, 1912. *Dinamika Atmosfery* (St.Petersburg). Demchinsky published weather forecasts for Russia, the United States, and Japan.

16. According to other sources, *Klimat* existed between 1901–1904.
17. On critics, see, for example, Griboedov, S. D. 1901. *Predskazania Demchinskogo pered Sudom Nauki i Faktov* (St. Petersburg: n.p.); Doroshevich, Vlas M. 1905. *Bezvremen'e* (Moscow: I.D. Sytin). On Mendeleev's comments, see Witte, Sergey Yu. 1994. Memories (*Vospominaniia*) (Tallin-Moscow: Skif-Aleks) Vol. 1, pp. 153–154. According to certain sources, his long-term meteorological forecasts were quite accurate.
18. Demchinsky, Nikolai A. 1900a. *Organizatsiya Khozyaistva v Severnykh Guberniyakh* (St.Petersburg).
19. Demchinsky Jr. presents a very interesting but unstudied figure in the history of Russian literature, civic thought, and science. He visited Lev Nikolaevich Tolstoy many times and wrote memoirs about these visits; he was a musician of some subtlety, the author of several librettos, and a close friend of Sergey Prokofiev, one of the twentieth century's outstanding composers; he was a journalist and littérateur, a great proponent of Russian symbolism; he was an original thinker and an expert in philosophy, *mysticism*, and *esotericism*; he was an outstanding chess player and wrote books on the philosophy of chess. And, finally, and most importantly from the point of view of this article, Boris Demchisky was a professional agronomist with primary interest in cereals.
20. TsGA Mosrvy. F. 228. Op. 3. D. 1703. L. 94–94b.
21. TsGA Mosrvy. F. 228. Op. 3. D. 1703. Ll. 12–70.
22. TsGA Mosrvy. F. 228. Op. 3. D. 1703. Ll. 14b, 23, 29, 31.
23. TsGA Mosrvy. F. 228. Op. 3. D. 1703. L. 84.
24. See Vishnevetskii, Igor'. 2009. *Sergey Prokof'ev* (Moscow: Molodaya gvardiya).
25. See Demchinsky, Boris N. 1908. *Rossiya v Man'chjurii (po neopublikovannym dokumentam)*, (St. Petersburg: n.p.).
26. Demchinsky N., 1900a; and idem. 1903. *Nujdy Sel'skogo Khozyaistva i Budushchee Rossii* (St. Petersburg: n. p.).
27. RGAE. F. 478. Op. 5. D. 2379. L. 207.
28. Demchinsky, Nikolay A., Demchinsky, Boris N. 1908. *Obespechennost' Urojaya. Teoriya i Praktika Gryadkovoi Kul'tury Khlebov* (St. Petersburg: n. p.).
29. Ibid, pp. 2–17.
30. Ibid, pp. 1–2.
31. Ibid, p. 92. It could be suggested that the journeys made by the Demchinskys to the East had some bearing on the development of

their method: Nikolai Aleksandrovich was the war correspondent of the *Novoe Vremya* newspaper during the Russo-Japanese war of 1905. His son Boris spent some time with his father in Manchuria also as a journalist. It can be assumed with some confidence that Demchinsky Jr., as a professional agronomist, also showed interest in the agriculture of the region, and its particular agrotechnical approaches.

32. Demchinsky N., Demchinsky B., 1908, pp. 3–12, 91–92.
33. Later, Boris did place articles about the BCCC in *Zhurnal zemledel'tsa* (The Farmer's Journal), a magazine that he published in St. Petersburg in 1912 and 1913.
34. Demchinsky N., Demchinsky B., 1908, p. 1.
35. See Prokof'ev, Sergey S. 1973. *Avtobiografiya* (Moscow: Sovetskii Kompozitor).
36. I came to this conclusion while studying the Collection of the Ministry of Agriculture in the Russian State Historical Archive (hereafter–RGIA): F. 381 (Office of the Ministry), F. 382 (Agricultural Scientific Committee), F. 398 (Department of Agriculture of the Ministry).
37. n.n., 1931a. "Demchinsky Nikolay Aleksandrovich." *Bol'shaya Sovetskaya Entsiklopedia*, Izd. I. (Moscow: Sovetskaya Entsiklopedia). Vol. 21, p.134.
38. For an overview of agriculture in the late Russian Empire see, for example, Klepikov, S.A. 1917. *Atlas po Agrarnomu Voprosu* (Moscow: n. p.).
39. For details, see Elina, Olga. 2002. "Planting Seeds for the Revolution: The Rise of Russian Agricultural Science, 1860–1920." *Science in Context*, 15 (2): 209–237; idem, 2008. *From the Tsar's Gardens to Soviet Fields: A History of Agricultural Experimental Institutions, XVIII c. to the 1920s.* (Moscow: Egmont-Russia). 2 vols.
40. Ermolov, Aleksey S. 1905 (1995) *Narodnoe Pogodovedenie* (Moscow: n.p.), p. 8.
41. Ermolov, Aleksey S. 1901–1905. *Narodnaya Sel'skokhozyaistvennaya Mudrost' v Poslovitsakh, Pogovorkakh i Primetakh.* 4 vols. (SPb: n. p.), vol. 4, p. 268.
42. See Anan'ich, Boris V., Ganelin, R. Sh. 1999. *Sergey Yul'evich Witte I ego Vremya* (SPb: Dmitriy Bulanin), p. 320.
43. Witte, 1994, vol. I, p. 155.
44. Anan'ich, Ganelin, 1999, pp. 320–321.

45. See *Deyateli Revolutsionnogo Dvijeniya v Rossii: Biobibliograficheskii Slovar'*. 1930. 5 volumes. (Moscow: Izd-vo Vsesoyuznigi Obshchestva Politicheskikh Katorjan i Ssyl'no-Poselentsev, 1927–1934). Vol. 2.
46. V. M. Bokova, ed. 2005. "Politicheskie proekty N. A. Demchinskogo," *Rossiiskii arkhiv*, 14: 605–642.
47. Demchinsky N., Demchinsky B., 1908, pp. 120–121.
48. Pisarjevsky, Oleg N. 1963. *Dmitriy Nikolaevich Pryanishnikov* (Moscow:), p. 154.
49. See "Otchet Novozybkovskogo Opytnogo Polya za 1908 g.," in *Otchet o Rabotakh Novozybkovskogo Opytnogo Polya, 1908–1910 gody* (Novozybkov: n. p.).
50. For the review of the farming conditions in Chernigov province, see Rusov, Aleksandr A. 1898–1899. *Opisanie Chernigovskoi Gubernii* (Chernigov: n.p.), 2 vols.
51. Lisitsyn, Petr I. 1908. "Nekotorye dannye o gryadkovoi kul'ture khlebov po Demchinskomu," *Vestnik sel'skogo khozjaistva*, 46: 15–21.
52. RGAE. F. 478. Op. 5. D. 2379. L. 208.
53. See Demtschinsky, Nikolay A., Demtschinsky, Boris N. 1909. *Die Vervielfachung und Sicherstellung der Ernteerträge. Theorie und Praxis der Ackerbeetkultur. Umgearbeitet nach der 6en russischen Auslage.* (Berlin: Verslagsbuchhandlung Paul Parey); Demtchinsky, Nikolay A., Demtchinsky, Boris N. 1911. *Die Ackerbeetkultur, ihre Grundlage, Methoden und neuesten praktischen Ergebnisse.* (Berlin: Verslagsbuchhandlung Paul Parey); Demtchinsky, Nikolay A., Demtchinsky, Boris N. 1913. *Méthode pour obtenir de forts rendements en céréales: exposé des procèdès de culture en billons.* (Paris: Librairie Chapelot).
54. See, for instance, n.n., 1912. "Recent Work in Agricultural Science—Field Crops," *Experiment Station Record*, 27: 231–233; Russell, E. J. 1919. "The Possibilities of Increased Crop Production," in *Life and Its Maintenance. A Symposium on Biological Problems of the Day* (London (UK); Bombay (India): Blackie and Son Ltd.), pp. 95–117; Soskin, S.E., no date. *The Bed Cultivation of Cereals: An Intensive Method for the Growing of Cereals* (London, UK: The Zionist Organization).
55. See Glover, Dominic. 2013. "Scientific Rice Cultivation: Historical Outlook," *Science in agriculture: Interactions of science, agriculture, and environment. XXIV ICHST*, (Manchester).

56. To this day it is recognized as one method of growing this crop. See Glover, 2013.
57. RGAE. F. 478. Op. 5. D. 2379. L. 206.
58. Ibid., L. 208.
59. Ibid., L. 208b.
60. Ibid., Ll. 207–207b.
61. Demchinsky, Boris N. 1923. *Vozmezdie za Kul'turu* (Leningrad), pp. 20–25.
62. RGAE. F. 478. Op. 5. D. 2379. Ll. 204–206.
63. RGAE. F. 478. Op. 5. D. 2379. L. 248.
64. Demchinsky, Boris N. 1931. *Peresadka Rastenii Polevoi Kul'tury* (Moscow-Leningrad: Gisizdat. Sel'khoz. i Kolkhozno-kooperativnoi Literatury); and idem, 1932. *Puti Razvitiya Sel'skokhozyaistvennogo Proizvodstva* (Moscow:).
65. Ibid, p. 78.
66. See Demchinsky, Boris N., Sumgin Mikhail I. 1940. *Oblast' Vechnoi Merlsoty* (Moscow-Leningrad).
67. Sapegin, Andrei A. 1935. "Znachenie iarovizatsii dlia fitoselektsii," in *Teoreticheskie osnovy selektsii rastenii* (Moscow-Leningrad: Gosizdat s. kh. Lit-ry), vol. 1, pp. 807–814; here, p. 814.
68. Roll-Hansen, Nils. 2005. *The Lysenko Effect. The Politics of Science* (Amberst, New York: Humanity Books).
69. Lysenko, Trofim D. 1928 *Vliyanie Termicheskogo Faktora na Prodolhzitel'nost' Faz Razvitiya Rasteniy. Opyt so Zlakami I Khlopchatnikom, Trudy Azerbaydhzanskoy Tsentral'noy Opytno-Selektsionnoy Stantsii im. Ordhzonikidze.* Baku. № 3; idem, 1935a. *Teoreticheskie Osnovy Yarovizatsii.*; and idem, 1935b. *Selektsiya i Teoriya Stadiynogo Razvitiya Rasteniy.*
70. Witte, 1994, vol. 1, p. 154–155.
71. Fedorovich, Vit. 1927, "Polia zimoi," *Pravda*, August 7, p. 5.
72. n. n., 1931b. "Redaktsionnoe soobshchenie," *Izvestiia*, October 29, p. 3.
73. See Rossianov, Kirill O. 1993. "Stalin as Lysenko's Editor: Reshaping Political Discourse in Soviet Science," *Configurations.* 1 (3): 439–456.
74. Vavilov, N.I. 1936. "Botanic-geographical background of plant breeding", in *Theoretical background of plant breeding* (M.-L.: Gosizdat Sel'khoz. Sovkhoznoi I Kolkhoznoi literatury), p.72–73.
75. Soyfer, Valery N. 1994. *Lysenko and the Tragedy of Soviet Science* (New Brunswick: Rutgers University Press), pp. 53–132.

76. Ibid, pp. 133, 134.
77. See, for example, Joravsky, 1986; Roll-Hansen, 2005.
78. For example, in his report for 1932, Lysenko analyzed the data sent in from the Southern Ukraine farms. In the Kharkov region, out of eight farms, two observed a decrease in yield; one reported no change; one saw an increase of 50 kg/ha; three up to 100 kg/ha; one claimed 150–200 kg/ha. In the Odessa region, three farms noted a decrease in yield; one reported no change; 11 claimed an increase of up to 50 kg/ha, while four reported an increase of up to 100 kg/ha. In the Donetsk region, there was a decrease in yield on five farms and no change on three farms, while on two farms, there was an increase of up to 50 kg/ha. See Lysenko, Trofim D. 1932. "Predvaritel'noe Soobshchenie o Yarovizirovannykh Posevakh Pshenits v Sovkhozakh i Kolkhozakh," *Byulleten' Yarovizatsii*, 2–3: 3–15.
79. See Lysenko, 1932; and idem, 1935c. "Ocherednye Zadachi Yarovizatsii," *Sotsialisticheskoe Zemledelie*, October 29. P. 3.
80. Lisitsyn, Petr I., et all, 1929. "Iarovizatsiia Ozimykh," *Sel'skokhoziais tvennaia Gazeta*, November 13 (№ 212). P. 3.
81. Fitzgerald, Deborah. 1996. "Blinded by Technology: American Agriculture in the Soviet Union, 1928–1932," *Agricultural History*, 70: 459–486.
82. For example, the Novo-Urenskaya station (in Ul'yanovsk region) tested several aspects of Lysenko's method—in particular, the influence of the length of vernalization on the plant's ability to resist cold. There were detailed trials comparing the viability of vernalized and nonvernalised seeds of different crops, with different lengths of vernalization. Scientists at the station recorded no improvement, and sometimes even negative effects from vernalization. Thus, when it comes to the results of trials on Lisitsyn's No.11 winter rye, the report reads: "At the beginning of harvesting, following a vernalization period of 60 days, not one single plant of this variety remained…"; whereas Tsezium 111, a variety of spring wheat: "does not react strongly to vernalisation and even offers a negative result" (Gurkin, 2011, pp.).
83. Stalin, Iosif V. 1934, "Otchetnyi doklad o rabote TsK VKP(b) XVII s"ezdu partii," in *Stenograficheskii otchet o rabote XVII s"eda VKP(b), 26 ianvaria-10 fevralia 1934 g.* (Moscow: Partizdat), p. 19.
84. Elina, 2008, vol. 2, pp. 238–239.

85. RGAE. F. 478. Op. 5. D. 2397. L. 176.
86. Lysenko, Trofim D. 1935a. *Teoreticheskie Osnovy Yarovizatsii*; idem, 1935c. Ocherednye Zadachi Yarovizatsii. *Sotsialisticheskoe Zemledelie*, October 29, 277.
87. Sheehan, Elena. 1985. *Marxism and the Philosophy of Science: A Critical History*. (Humanities Press).
88. I have to add that it was *just a method of vernalization* that was evaluated—but not its originator, and, of course, not all the events that followed the appearance of Lysenko on the stage of Soviet agriculture. Therefore, the experts were not responsible for Lysenko's further "panaceas."
89. For the details, see Kalashnikov, P. A. et al. 2012. "Grebnevaia tekhnologiia vozdelyvaniia pshenitsy v usloviiakh oroshaemogo zemledeliia," at http: //www.rusnauka.com/25_PNR_2012/Agricole/1_115934.doc.htm; Karabalaeva A. D. Sydyk, D. A. 2012. "Grebnevaia tekhnologiia vozdelyvaniia ozimoi pshenitsy v usloviiakh orosheniia Iuzhnogo Kazakhstana." http://borona.net/hight-technologies/chemicals/Grebnevaja_tehnologija_vozdelyvanija_ozimoj_pshenicy_v_uslovijah_oroshenija_uzhnogo_Kazahstana.html; Ospanbaev, Zh. O. 2000. "Rezul'taty ispytaniia grebnevoi tekhnologii vozdelyvaniia pshenitsy na iuge i iugo-vostoke Kazakhstana," *Semenovodstvo i selektsiia pshenitsy v Tsentral'noi Azii*, (Almaty). 2(5): 1–21; Sayre K.D., Moreno Ramos O.K. 1997. "Application of Raised-Bed Planting System to Wheat," *WPSR* No 31 (12) (Mexico: CIMMYT); Aquino P. 1998. "The Adoption of Bad Planting of Wheat in the Yagui Valley, Sonora, Mexico." *WPSR* No 17a (3). Mexico: CIMMYT; Freeman, K. W., et al. 2008. "Bed and Flat Planted Winter Wheat as Influenced by Row Configuration," at http://www.nue.okstate.edu/Index_Publications/Bed_Flat_Planted_Wheat.htm

State Officials and Would-Be Scientists: How the Ukrainian Ministry of Agriculture Discovered for Lysenko that He Had Made a Scientific Discovery

Lukas Joos

On October 8, 1929, *Pravda*, the main newspaper of the Soviet Communist Party, carried a lengthy article with the title "On the Sowing of Winter Cultures in Spring (The Discovery of Agronomist T. D. Lysenko)".[1] Its author was Aleksandr Shlikhter (1868–1940), the head of the Ukrainian Ministry of Agriculture [NKZS].[2] In contrast to the rather sober title, the article's text was permeated with praise for Trofim Lysenko's (1898–1976) discovery. This discovery of "exceptional scientific and economic value," Shlikhter explained, would lead to Soviet grain yields becoming higher and more stable already in the immediate future; its main benefit was that it would put an end to grain losses caused by Black Frost.[3]

Some industrial crops exhibit a winter and a spring form; Black Frost affects the higher-yielding winter forms. The seeds of winter crops are sown in autumn; the plants ripen the following year (if they are, like spring cultures, sown in spring, they do not ripe and produce no yield). Black Frost occurs in extremely cold winters with little snow: without a sufficient snow cover, the soil can get cold enough to kill the seeds laying in it.

The remedy against Black Frost that Lysenko allegedly had discovered[4] in early 1929 consisted of a method to make winter cultures ripen and

L. Joos (✉)
University of Zurich, Zurich, Switzerland

W. deJong-Lambert, N. Krementsov (eds.),
The Lysenko Controversy as a Global Phenomenon, Volume 1,
DOI 10.1007/978-3-319-39176-2_3

produce yield even when sown in spring. According to Shlikhter, Lysenko had discovered that: (1) The seeds of cereals go through a certain early stage in their development during which they need only a certain humidity and low temperatures; (2) It was possible to artificially produce these conditions; that is, seeds can be made to pass this stage without sowing them; (3) An easy and practical way to do this was to first germinate them slightly and then keep them for a certain number of days in the snow.[5]

At the end of the 1920s, Black Frost was the worst weather-related problem in Ukrainian agriculture. Two times in a row, during the winters of 1927–1928 and 1928–1929, large portions of winter wheat had succumbed to it. This had led to major crop failures in a country where, due to technological backwardness, grain yields were significantly lower and more unstable than in Western Europe or America.[6] A grain crisis was the result. In summer 1928, the Ukrainian State Commission for Aid to Victims of the Crop Failure was formed, with Shlikhter as its head. The commission engaged the central Soviet government for help and organized food and fodder relief for the most severely impacted areas.[7] Also in summer 1928, a large scientific event to address the problem of Black Frost was organized. Yet the "establishment" of Soviet plant sciences was unable to present a solution.[8] The Black Frost crop failures and Lysenko's alleged discovering of a solution to it took place during the eve of collectivization. Stalin had set the course for collectivization at the XV party congress of December 1927.[9] During 1928, Stalin and his confidants subdued those within the party and the state administration who opposed collectivization. While since early 1928, political pressure was increasing on all actors of the (predominantly "bourgeois") Soviet science community, agricultural scientists, which had opposed collectivization virtually unanimously, were particularly targeted.[10] After the two violent grain requisition campaigns of early 1928 and early 1929, Stalin and his confidants decided in autumn 1929 that collectivization was to be carried out without delay and at any rate.[11]

In December 1929, when the Soviet countryside began to submerge in violence and chaos, the *Agricultural Gazette* [SG][12] carried an article[13] signed by the NKZS. Its purpose was to defend Lysenko against a number of "bourgeois" plant scientists who, a month earlier and also on the pages of the *SG*, had failed to match Shlikhter's level of enthusiasm about *vernalization*, as Lysenko's discovery was dubbed.[14] For Lysenko, this article was the perfect finale of a six-month-long period that had started in July, when vernalization first hit the news, and propelled him from obscurity to the limelight of Soviet agricultural debates. First, vernalization was discussed in the central press, mainly by officials from the NKZS and newspaper reporters. Wild enthusiasm dominated this coverage. Then, in October, the discussion migrated to the *SG*. After the publication of the scientists' opinions in November, Lysenko received ample room to explain how the scientists had underestimated his work in early December. A little later, the NKZS's defense of Lysenko was published. In this defense, the ministry resorted to political slander of the scientists. This changed the "ground rules" of the discussion of Lysenko and vernalization, which up to that point had been purely scientific.

In addition to quick fame, the "discovery" of vernalization made Lysenko's scientific career skyrocket. From a junior rank-and-file position at an experimental plant-breeding station in Gandzha, a town in the remote Azerbaijan SSR, he was elevated to head of his own research group at the Ukrainian Institute for Genetics and Plant-Breeding (UGSI)[15] in Odessa, the most important agricultural research center of the Ukrainian SSR.[16] This transfer to Odessa, which Shlikhter had ordered by a special decree, gave Lysenko the possibility—in the words of Valery Soyfer—to "skip many steps" of a usual scientific career "at once."[17]

There was, however, a downside to the story—not for Lysenko, but for Soviet agriculture. As a tool to raise crop yields, vernalization was worthless. Its use during the 1930s resulted in nothing but wasted resources.[18] Thus, from a scientific point of view, there were no reasons whatsoever for the praise and promotion Lysenko had received for it.

In this chapter, I will examine the reasons for this unsubstantiated promotion. In previous scholarship, this promotion has often been explained in the way later coups of Lysenko were explained. Lysenko would come up with an idea that was clearly pseudoscientific, and therefore, beyond the scope of scientific discussion.[19] Nevertheless, by instrumentalizing the Stalinist propaganda machine to exert political pressure, he would then succeed to have it implemented. It is held that in the 1929 promotion of vernalization, Lysenko and the newspapers, which propagandized his program, played the active part. The other actors—that is, NKZS and the plant scientists—were pressured to react to what Lysenko and the press did and said. The NKZS, in particular, is believed to have been in an extraordinarily difficult political-administrative situation, as the party elite would have expected it to resolve the problem with Black Frost quickly. When Lysenko started to advertise his solution in bold terms, the NKZS, it is argued, had no real choice but to endorse vernalization uncritically.[20]

Further, it is held that the scientists who were involved in the discussion encountered similar problems. To call vernalization the pseudoscientific nonsense that it was would have involved a deliberate risk of getting fired, arrested, or even shot.[21] More recently, Nils Roll-Hansen has shown that at the end of the 1920s, Lysenko was not a pseudoscientist yet and that some aspects of his early work in plant physiology even received some positive scholarly attention.[22] Thus, the value of vernalization was a scientifically legitimate question. The reason for its overly positive evaluation, Roll-Hansen holds, was the failure of the NKZS, the press and even some of the scientists to properly discern between the different evaluation criteria that apply for basic and applied science. Because of this failure, it never occurred to them that vernalization might be useless in practice even if it worked in laboratory conditions.[23]

In this chapter, I will argue that in 1929, the NKZS was not only the main promoter of Lysenko but also the main responsible party for the overly positive assessment of his "discovery." My argument is based on a close analysis of newspaper articles published mostly between July 21 and December 19, 1929. Introducing a small number of previously unknown, published Ukrainian language sources that I was able to find in Moscow libraries,[24] I will demonstrate that the journalists simply echoed the NKZS officials, who ignored not only scientists' call for caution but also reservations expressed by Lysenko himself. The reason for the NKZS flawed actions, I hold, was not political pressure, but scientific mismanagement, combined with the ministry's political-administrative power to make authoritative statements and decisions on specific scientific issues. This claim is based on strong circumstantial, but not on archival evidence. Therefore, it must be regarded as preliminary.

The structure of this chapter is simple. A brief overview of the scientific context of Lysenko's "discovery" is followed by the analysis of the relevant sources in chronological order. In the third section, the findings from the analysis are evaluated and the roles that the four main groups of actors—the NKZS, Trofim Lysenko and his father Denis, the journalists, and the plant scientists—played for Lysenko's career leap are discussed.

The Scientific Context of Lysenko's "Discovery"

In January 1929, Lysenko was one of over a thousand foreign and Soviet scientists to participate in the First All-Union Congress on Genetics, Plant Breeding, Seed Production and Livestock Breeding, organized by the eminent plant scientist and "patron" of Soviet plant sciences Nikolaĭ Vavilov (1887–1943). Lysenko submitted a paper, "On the Question of the Nature of Winter Cultures,"[25] coauthored by Donat Dolgushin (1903–1995), a coworker from Gandzha who later became his assistant. In a preliminary discussion, Vavilov[26] and the head of the plant physiology department of his All-Union Institute of Applied Botany and New [Plant] Cultures [VIPBiNK],[27] Nikolaĭ Maksimov (1880–1952), agreed it was weak. But while Maksimov did not want to put it on the agenda, Vavilov insisted that presenting a paper at such an event would be a good learning experience for Lysenko.[28]

The subject of Lysenko's paper was the effect of low temperatures on certain developmental stages of winter crop cultures. In the 1920s, the possibility to affect plant development by changes of light and temperature regimes was a relatively new research topic, believed to have practical implications for agriculture. In 1918, German botanist Gustav Gassner (1881–1955) had published a pioneering study arguing that the flowering of winter annual plants was dependent on a period of low temperature in their early stages of development. During the 1920s, Maksimov continued Gassner's experiments in laboratory and greenhouse conditions. He referred to these experiments as *kholodnoe prorashchivanie* (cold germination). Unlike Lysenko, Maksimov exposed plant seedlings, not seed, to low temperatures. This, of course, excluded the possibility of the transfer of these experiments into agricultural conditions.[29]

Lysenko's early work was influenced by the widespread belief that plants need a certain "sum of heat" to complete certain early stages of their development. The aim of a larger publication of Lysenko that appeared in 1928[30] was to discover the "simple and fundamental" law of the dependence of plant development on temperature. This publication contained a mathematical formula to calculate the "cold requirement" plants allegedly had.[31]

The aim of the experiment that Lysenko described at the January congress was to develop further quantitative aspects of his theory on cold requirement of winter crops. In early 1928, Lysenko divided the seed of winter barley and winter wheat into three batches. Two batches he soaked in water (the one for a shorter, the other for a longer period of time); the third served as a control. Then, he put the seed in sacks and kept them in the snow. Beginning from March 2, sowings were carried out every few days. From his observations of how the plants headed, Lysenko, according to Roll-Hansen,[32] concluded that "once germinated or swelled [i.e., soaked in water] seed had received the necessary period of low temperature, the plants would spike whenever they were sown".[33]

As Lysenko argued, the results of this experiment showed that low temperatures had a greater effect on early development stages of winter crop cultures than day length.[34] He saw his "general theory of stages" confirmed: plant development and growth were to be separated. The first stage of development required low temperatures; the completion of the following stages, however, would accelerate with higher temperature. Note that Lysenko did not discuss the practical implications of his work.

Maksimov criticized Lysenko sharply. He argued that Lysenko's argument was based on too few experiments, which would never generate data accurate enough to be quantifiable. He pointed out that the fact that one could sow winter cultures in spring by exposing them first to a cold treatment had been discovered a century before and that Lysenko's theoretical conclusions lacked explanatory value. The belief that the main difference between winter and spring wheat was that they needed different quantities of low temperatures during their early development was erroneous.[35] Lysenko was not mentioned in the extensive media coverage of the congress, and he returned from Leningrad to Gandzha the rank-and-file scientific worker of a remote experimental station he was when he arrived there.

The Discovery of a Discovery: The Promotion of Vernalization from July to December 1929

According to Shlikhter's October account, Lysenko's father Denis showed up at the NKZS on May 10, 1929, and demonstrated to the officials samples of winter wheat. He stated that he had sown this wheat in spring, after treating it with his son's 'method.[36] As a result, on July 12–13, a commission of the ministry inspected his farm and determined that the winter wheat had grown exceptionally well and would yield approximately three times as much grain as the local spring sorts. A week later, on July 19, Trofim gave a speech at a special meeting of the NKZS board,[37] also attended by representatives of experimental stations.[38]

Two days later, on July 21, the story, "The Discovery of the Agronomist Lysenko," appeared in the *Pravda*.[39] It contained a report by a journalist, Vl. Grigor'ev, and a statement of a certain Gorban',[40] a deputy head of the NKZS. Grigor'ev informed readers that at a meeting of seed producers, NKZS representatives had made a statement on the "practical results" of Lysenko's work on the winter wheat "Ukrainka." Without clarifying the content of the statement, Grigor'ev went on to describe the field experiment and the subsequent inspection by the commission.[41] He stated that the "perspectives that ensue from this exceptional discovery of the agronomist Lysenko, which is confirmed by so splendid experimental data," were "so great that they cannot be effectively counted right away,"[42] but did not mention any data or even explain what exactly had been discovered. The commission's estimate of the yield, Grigor'ev explained, "should be expressed [*sic*] approximately around [*sic*] 3 tons per hectare."[43] The grammatical and semantic errors contrasted with Grigor'ev's effort to imitate Soviet officialese. Like Gr igor'ev, Gorban' did not explain what Lysenko's discovery consisted of, but emphasized its importance: it was of "absolutely exceptional value"; its implementation would play an "enormous role," and it would "be applied in practice in the sovkhozy and kolkhozy of the Ukraine." Strangely enough, his statement ended with the observation that "if Lysenko's method proves its worth, it will have incalculable consequences for the country's entire agricultural sector."[44]

On July 27, a week after the initial report to the NKZS board, both Trofim and Denis Lysenko gave a report at an event in Kharkov. On that occasion, they gave the *Komunist*[45] an interview. Denis told the reporter that on his way to the January congress, his son had visited him in his native village of Karlovka. During this visit, Trofim had "acquainted" him with his seed treatment.[46] When he began to suspect that the autumn sowings on his farm would be killed by Black Frost, he decided to give his son's "method" a try. The detailed description of how he first soaked and then stored the seed in the snow makes it clear that he used exactly the technique that his son had described at the January congress.[47]

Trofim's interview appeared under the headline "A Valuable Invention in the Field of Agriculture," and the subtitle "Is it Possible to Sow Winter Cultures in Spring?"[48] Note that this question was part of the title which also stated that a valuable discovery had been made and that it was printed on the same page as Denis' account of his success with sowing winter cultures in spring. Trofim presented himself as a researcher interested in theoretical matters. He said that he "came upon the question of the 'winterness' [ozimost'] of plants" while he "was studying the matter of the dependence of the development stages of plants from temperature."[49] He added that after a period of chilling, any sort of winter bread crops would ripen if sown in spring. He explained that this chilling technique would allow speeding up the development of both winter and spring cultures. Note that Lysenko claimed to be the inventor only of the theoretical concept of "winterness." The seed chilling technique and its implications for speeding up plant development, he attributed to his workplace. Note also that he used correct terminology to describe the seed chilling (in Ukrainian: *okholodzhennia*).[50]

Of course, the reporter wanted to know about the agricultural implications that the findings of the Gandzha plant-breeding station would have. Lysenko gave a crystal-clear answer:

> It is clear [...] that it will be necessary to verify all this laboratory research and the speculations (*mirkuvannia*) on the practicability of the new method of sowing winter and spring varieties by practical means. The Narkomzem of the Ukraine [i.e., the NKZS] starts such verification on its experimental stations in the near future. *Only after such verification it will be possible to speak with certainty* (pevno) *about the practical application of this theoretical scientific work, which the Azerbaijan agricultural experimental station has carried out* [italics are added—L. J.].[51]

The absurdity of the evolving situation should be clear. Lysenko's father carried out a simplified version of the experiment his son had described in detail at the January congress in front of a large scientific audience, without creating any sensation. Within a week after sending a commission to the Lysenko farm, a top official from the NKZS claimed that this experiment was an exciting, extremely valuable discovery which would have a major impact on agriculture. However, he could not tell whether the implementation of this discovery would prove worthwhile. A week later, Lysenko, never mentioning a discovery, explained that he had no idea about the practical usefulness of sowing winter cultures in spring, and that basically, it was useless to speculate on that question until further tests would be carried out by the NKZS.

On August 4, a week after the meeting in Kharkov, the newspaper *Economic Life* (ĖZh)[52] published a set of five (!) articles under the general heading "The discovery of agronomist Lysenko." The first article was written by the newspaper's "correspondent in Kharkov." The other four were written by two senior officials of the NKZS and two directors of experimental plant breeding stations.

The correspondent's unsigned article, titled "The Discovery of agronomist Lysenko: Its essence," informed readers of the possibility of protecting winter cultures from Black Frost and spring cultures from the destructive influence of hot and dry summers.[53] The correspondent's explanations about the essence of the discovery were too confused to convey a clear opinion. He seemed to think that the exciting part was the seed chilling technique; the theory of plant development that it was based on he called "simple premises." Overall, the correspondent was no less enthusiastic than Grigor'ev had been on July 21; also, his grammar was slightly better than that of his colleague from *Pravda*.

The second article, "Insurance of the Winter Cultures against Freezing," was written by a certain Kachinskiĭ, another deputy head of the NKZS.[54] As Gorban', Kachinskiĭ was enthusiastic. But his article made it obvious that he was not only completely ignorant of the scientific context of Lysenko's "discovery," but had also remarkable troubles with sound reasoning. He referred to Black Frost (*vymerzanie*) as *vymorazhivanie*, which denotes a *willfully induced* freezing process. By adding that Lysenko's promising solution to Black Frost consisted in freezing (*zamorazhivanie*)[55] of the seed, he—without intention, of course—stated that the solution to Black Frost was basically Black Frost itself.[56] In Kachinskiĭ's opinion, the NKZS should "give a definite program of tasks" to all the experimental plant-breeding stations."

The third article, "More Research is Needed," written by Mashura, head of the NKZS experimental department, was also as positive as it was poorly written.[57] The last two articles were authored by the director of the Kiev territorial experimental station, Filipovskiĭ,[58] and the director of the Poltava experimental station, Sazanov. Filippovskiĭ echoed the praise of the NKZS and the journalists. Then, he pointed out that in order to carry out Lysenko's experiments, his station would need additional equipment. The initial part of Sazanov's article consisted of some remarks about Lysenko's discovery of "the effect of temperature on the 'winterness'" of plants, which would be of great value for the country's agriculture. However, two things set it apart. One, Sazanov made it clear that "the further success and useful application" of Lysenko's work would depend on a suitable method for chilling seed and believed the NKZS should hold a competition to develop it; two, it was written in literate Russian.

Contradictions between the articles are notable. The *ĖZh* correspondent thought that Lysenko's merit lay not in some theory, but in the possibility of exposing seed to an artificial cold treatment. Sazanov stated precisely the opposite. Neither of the NKZS officials mentioned Lysenko's statement that the ministry needed to do further tests before it could say anything definite on the agricultural value of his method or expressed any form of uncertainty or doubt about its effectiveness. The NKZS deputy explained the seed chilling not just incorrectly, but also differently than Lysenko, and the accounts of the two representatives from NKZS contradicted each other on the role of Lysenko's father. Mashura, head of the NKZS's *experimental* department, contradicted Lysenko's claim that further tests were necessary insofar as he stated that what would work in scientific conditions should be implemented into agriculture, and the statement from Kachinskiĭ made it clear that the top echelon of the NKZS saw nothing wrong in defining specific research programs, that is, commanding the research of its plant-breeding stations.

On September 1, Lysenko gave a report on "The Question of the 'Winterness' of Plants" at the VIPBiNK. Those present included Maksimov, V.G. Batyrenko[59] and the plant-breeder Viktor Talanov (1871–1936), a senior member of the VIPBiNK.[60] The protocol mentions that "In 1928,[61] a spring sowing of winter wheat in agricultural conditions was carried out in the Ukraine," and the yield was "utterly splendid"—so splendid, in fact, that Lysenko "*found it difficult to make conclusions of any kind* [Italics added—L. J.]" from it. "Probably," he pointed out, "some particularly lucky circumstances" had played a role.[62]

In the subsequent discussion, Maksimov said that Lysenko's research was interesting to "the utmost degree." However, he also noted that the method of chilling seed in the snow was invented by the horticulturist Grachev in the 1880s.[63] Lysenko's achievement was finding a "method for a quantitative determination of the winterness of plants, measured in days." Lysenko responded that he did not claim to be the inventor of the seed chilling method.

Talanov emphasized the contradiction between Lysenko's "unusual circumspection and humbleness" and the "extraordinarily practical value and theoretical interest" of his research. His suggestions included the following: Lysenko should write a report for publication in the journal of the VIPBiNK.[64] He should be invited to work at the institute, where his experiments should be repeated at the institute's laboratory, with him personally participating (*pri lichnom uchastii*). According to the protocol, no one spoke of a discovery; however, no one also demanded Lysenko explain how his experiments proved that crops only differed by their individual degree of "winterness." Less than two months later, Lysenko took up work at the Odessa institute, which was affiliated with the NKZS.[65]

On October 8, *Pravda* printed "On the Sowing of Winter Cultures in Spring," Shlikhter's eighteen-paragraph-long assessment of Lysenko's work. Sixteen paragraphs were written in the tone of unconditional praise. As Gorban', on July 21, Shlikhter emphasized the "extraordinary scientific and economic value" of the "discovery." He added that the fact that its practical implementation was "not difficult at all and fully accessible to mass peasant farming" made its value even greater. He did not indicate on what grounds he had come to that conclusion, and he did not mention the reservations Lysenko had expressed to the reporter of the *Komunist.*

In the two remaining paragraphs, he pointed out that further tests, experiments, and research were necessary. The NKZS was now busy planning these tests, which were scheduled for spring 1930. For the farms that would carry out test sowings, the NKZS would organize special trainings in the near future. Note that in terms of the organization of these tests, Shlikhter did not mention any form of cooperation with any scientific institute.

Shlikhter had trouble defining what Lysenko's "discovery" consisted in. He gave two, contradictory definitions. First, he explained that "[t]he point of the matter of this discovery consists in the following: it is established by the agronomist Lysenko that all grasses and a number of other plants have a certain period of development during which they need only a certain humidity and low temperature (the 'winterness' of plants)." Then, he stated that "[t]he discovery of Lysenko consists in that it is possible to produce artificially the period of low temperatures that is necessary for the seed of the winter cultures [...], without sowing them [the seeds] in the field." It should be clear that these two definitions describe two unrelated discoveries—one in basic and the other in applied science. The following, detailed description of Denis Lysenko's field experiment strongly suggests that Shlikhter thought that this experiment, too, was an element of the "discovery."[66]

After Shlikhter's assessment, the discussion migrated to the *SG*, which on October 10 published "Vernalization of Winter Cultures"[67] by L. Emel'ianenko of the NKZS.[68] To the best of my knowledge, Emel'ianenko's article is the first written account in which the term "vernalization" is used. Emel'ianenko set himself apart from his superiors by expressing himself in literate Russian and by not using the term "discovery." The article's first part consisted of an overview of recent research on why winter cultures do not flower if they are sown in spring. According to Emel'ianenko, a lot of bright minds—including Maksimov—had been trying to tackle the problem, but had failed to make any breakthrough. Now, Lysenko had made a "huge step forward in this regard." Emel'ianenko concluded that the NKZS was organizing "further testing" at experimental stations and state farms. No form of cooperation with any scientific institute was mentioned.

A month later, the NKZS formally adopted specific actions to be taken in connection with Lysenko's "discovery." The November 8–15 issue of the ministry's bulletin contained an instruction for test sowings "according to Lysenko's method," signed by Emel'ianenko, on behalf of the head of the ministry's department for field-husbandry. This instruction is notable in terms of form and content. It was exceptional in that it addressed a scientific problem, and in that most of the orders and instructions were signed either by one of the ministry's deputies or a department head. It stated that the Ukrainian Council of People's Commissars (i.e., Ministers) had approved test sowings on an area of 10,000 hectares throughout the Ukraine to be carried out in spring 1930.[69] A brochure was to be printed shortly and trainings for agronomists were to be organized. The instruction did not stipulate any cooperation with a scientific institute.[70]

That same week, the *SG* printed a series of four articles on Lysenko's vernalization, soliciting the opinion of "our eminent experts" on the matter.[71] Both the plant breeder Petr Lisitsyn (1887–1947) and the soil scientist Nikolaĭ Tulaĭkov (1875–1938) were renowned in their fields. They confirmed in vague terms that Lysenko's work had its merits and should be researched further, but from their assessments, it was not hard to get that they believed it had hardly any value for agriculture. Lisitsyn reminded readers that the effect of low temperatures on plant development was long known to plant physiologists. He also noted that factors such as day length, "that may seem insignificant," also had an effect on plant development.

In terms of practical importance, Lisitsyn said the main problem was the method was not technically suited for mechanized farming, while Tulaĭkov pointed out that the main advantage winter cultures had over spring cultures was their root system, which developed during the winter. If they were sown in spring, the [vernalized] winter cultures would not be higher-yielding than average spring cultures. The problem of Black Frost would be addressed more effectively by breeding varieties more resistant to frost and winter conditions.

Another expert, a certain professor Prik, pointed out the same problem as Tulaĭkov: if winter cultures were sown in spring, they would give lower yields. But unlike Tulaĭkov, Prik saw limited practical use in Lysenko's research. One thing, however, he made very clear: the amount of yield Lysenko's father had received from his field experiment was an exception made possible by unusually favorable climatic conditions.

The geneticist Andreĭ Sapegin (1883–1946), Lysenko's new boss in Odessa, stated, like Lisitsyn, that Lysenko had not come up with anything fundamentally new and that it was too early to think about practical implementations. Yet he presented Lysenko's theory of degree-days as original and useful. He informed readers that Lysenko would continue his research in Odessa and concluded that "one has to welcome the decision of the NKZU [i.e., the NKZS]" to organize further testing of the method. Note that he was the only scientist who commented on the actions of the ministry.

On November 19, the *SG* published two more articles and a brief commentary from its editorial staff. In "The Spring Sowing of Winter Cultures Opens Broad Perspectives," Maksimov pointed out that Lysenko had made no theoretical discovery; the theory behind Gassner's cold germination and Lysenko's vernalization was identical. The only difference was execution. Maksimov added that he and Gassner had thought that cold germination was too complicated to apply in practice; thanks to Lysenko, it now was available to the ordinary peasant. That this was a "first-rate achievement" was undeniable. However, Maksimov then pointed out that "Lysenko had a predecessor"—the horticulturist Grachev, whom Maksimov had mentioned during the September 1 meeting. Lysenko's discovery that the seed of different plants must be kept for different amounts of time at low temperatures was "extraordinarily valuable." However, Lysenko's contribution was clearly overestimated "by some circles." Maksimov did not state whom he had in mind by "those circles"; he simply expressed his hope that "their exaggerated expectations" would not impede sober assessment later. This was not the only criticism Maksimov leveled at those less knowledgeable in plant sciences than himself. He emphasized that freezing the seed (*promorazhivanie*)[72] would damage it; what was needed was chilling (*okhlazhdenie*).[73] Note that Maksimov's more sober and critical remarks conflict at least to some degree with the "extraordinary value" that Lysenkko's "first-rate achievement" allegedly had.

The other article, "Place the Verification Experiments of Lysenko's Method on a Wide Footing," by a certain V.T. Batyrenko—who had been present at the VIPBiNK-meeting of September 1—mixed possible problems with Lysenko's work, bold assertions of its enormous practical significance, and statements on the necessity to organize further testing. It is important to note that Batyrenko stated that results of Lysenko's father's experiment "we tend to consider as an exception,"[74] and that his statement that Lysenko's method consisted in "freezing" (*promorazhivanie*) seeds appeared on the same page as Maksimov's explanation that freezing would be counter-productive.[75]

The short comment of the *SG* editorial staff, "The Vernalization of Winter Cultures: A New Breakthrough in the Battle for Harvest," was—in terms of both its content and its language—very similar to the articles of the *Pravda* and the *ĖZh* correspondents. It consisted of overly long sentences, written in a formal tone, but with improper grammar; the content was confused. Probably in an effort to give a balanced comment, the editorial staff lumped together a series of statements evidently taken from various articles, critical and enthusiastic alike. The comment stated, for example, that cold germination was already well-researched and that vernalization would not lead to fundamental changes in agriculture, but also pointed out that Lysenko had found a "radical way" to evade the problem with Black Frost and resolve the "grain problem," and that grain yields overall would rise.[76]

On December 7, the *SG* featured a long treatise by Lysenko himself titled "What is the Essence of the Hypothesis[77] of the 'Winterness' of Plants?"[78] Lysenko's tone reveals that the scientists' articles had upset him. Particularly Maksimov's statement that he had made no theoretical discovery seemed to have seriously offended him. The stated aim of his treatise was to clarify how different the theory behind Gassner's cold germination was from his hypothesis of winterness. In an apparent attempt to correct Maksimov's mistaken perception of himself as a mere practitioner, Lysenko spent a lot of ink comparing his own approach with cold germination.[79] He pointed out that in Gandzha, practical applications of his theories had not interested him yet. Apparently to set himself apart from other practitioners, he even resorted to the quite extraordinary strategy of playing down the practical value others had attributed to his work. He argued that his winterness-method was more complex than cold germination, and thus less suitable for agriculture:

> At any small or large farm, it is incomparably simpler, easier and more convenient to germinate the seed in the cold, at 2–5 degrees [i.e., what, according to Lysenko, cold germination called for] than to keep the seed—after germinating it in the warmth—in the snow during almost 60 days. The most difficult [part] of the winterness-method is precisely the storing: one needs to see to it that the seeds do not grow further and, at the same time, are not killed by frost."[80]

This statement, however, did not prevent him from making other, boastful claims about the benefits of his work for Soviet agriculture. He now flatly contended that the results of his father's experiment demonstrated the "great practical value" of his method. While he conceded that a suitable way to use vernalization in agriculture had not been fully developed yet, his earlier uncertainty on whether the implementation of vernalization into agriculture would make sense had disappeared. Instead of pointing out the need for further testing, he now announced that the possibility of sowing winter cultures in spring was only a minor element of what his hypothesis of winterness would give Soviet agriculture.[81] Note that while Lysenko made rude, inappropriate comments about Maksimov's scientific competency and called him "his opponent," he did not accuse him of being disingenuous.

A few days later, the *SG* published the ministry's defense of Lysenko.[82] The official first gave a schoolmasterly assessment on how well the plant scientists had acquainted themselves with Lysenko's work. He concluded that only Sapegin had exhibited satisfactory knowledge. Lisitsyn, Prik and Tulaĭkov were criticized sharply. The official introduced political criteria to evaluate Lysenko's work.[83] On the one hand, he stated, for example, that the "hypothesis of Lysenko" was "splendid" because it provided the "nuts and screws" which were needed for Socialist construction. On the other hand, he deliberately misquoted Prik and insinuated that Lisitsyn had pushed the *SG* to conspire against Lysenko.[84] In terms of the specific scientific points which the scientists had raised, the official had little to say: the scientists were wrong, Lysenko was right, end of story. In any case, the official's language, grammar, and way of reasoning reveal that he must have lacked even a proper secondary education. Thus, even if he had wanted to, he would not have been able to address the scientists' criticisms from a scientific point of view. The official, however, made several clumsy attempts to demonstrate that the NKZS had been well aware of the difficulties with the implementation of Lysenko's "discovery," and knew the science behind it.[85] Below the official's article, the editorial staff of the *SG* published the following announcement: "We are attaching exceptionally great importance to the experiments of comrade Lysenko and confirm the actual necessity of taking every measure to ensure the success of his work."

Before we assess the roles different actors played in the promotion of Lysenko and his work during the second half of 1929, let us see how the two responses to the scientists' evaluations fit into the framework of the discussion. First, Lysenko, obviously, was a lot bolder than he had been six months earlier. Contrary to what he had told the *Komunist* reporter in July and the scientists at the VIPBiNK in September, his father's field experiment now had explanatory value. Second, Lysenko contradicted Shlikhter by stating that his method was comparatively difficult to use in agriculture and that a practical way to implement it had not been developed yet. The response from the NKZS demonstrated that the ministry saw nothing wrong with an agricultural official reassessing the assessments of some of the country's leading scientists.

Discoverers and Discoverees: Roles and Reasons

It seems fairly obvious that there was no organized media campaign to promote Lysenko independently from the NKZS. The ministry's officials were the ones being quoted. When the NKZS accused the *SG* of being influenced by Lisitsyn, the newspaper felt obligated to publicly assure its alignment with the ministry. It was not the NKZS that had to adhere to the media's viewpoint. Much rather, the media adhered to the one of the ministry. The scale of the NKZS's administrative blunder in reacting to news of the events is clear. Given that Lysenko had stated in July that without additional tests by the NKZS, it would be impossible to reach any definite conclusion, and given that the NKZS did not order such tests till November, the euphoria Shlikhter expressed in October is evidence of a total failure to observe even the most basic standards of scientific reasoning and methodology.

Biographical sources yield further evidence on Shlikhter's scientific ill-preparedness to assess Lysenko's work and administer a sound verification. He graduated from a rural *gimnaziia* as an external. On his only attempt to obtain a higher education (in medicine), he gave up in 1891, having enrolled just a year earlier.[86] I was not able to find biographical data on Gorban' and Kachinskiĭ. However, judging by their "oeuvres" on Lysenko's "discovery," I believe it is safe to suppose that to the extent that Shlikhter involved his deputies in his decision-making, he worsened things further. Moreover, the fact that everyone, apart from Emel'ianenko, kept referring to Lysenko's work as a discovery shows that they must have been ignorant not only of contemporary research in plant physiology but also of the elements that constitute a scientific discovery. This ignorance, in turn, must have contributed significantly to their impression that Lysenko's father's lucky strike was compelling evidence of something revolutionary. This belief would explain why the ministry never provided experimental data to support their enthusiasm and, once confronted with the scientists' evaluations, made a desperate attempt to appear well-informed and knowledgeable in December. It seems that it never occurred to anyone of the ministry's top echelon that the results from the field experiment might not be what they thought.[87]

Since the NKZS officials played the most decisive role in elevating Lysenko's status, it is important to consider their situation when they first heard about his father's experiment. Theoretically, four different situations are possible: (1) They were under significant pressure from the party bosses to find a solution to Black Frost and had significant doubts about the value of Lysenko's "discovery." (2) They were under administrative pressure and had no doubts. (3) They were under no pressure but had doubts. (4) They were under no pressure and had no doubts.

The third can be excluded right away. As mentioned, it has often been argued that the main reason for the flawed decision-making of the NKZS officials was that they were either in situation (1) or (2). If we assume (1), we run into difficulties. If we assume that the officials decided to implement vernalization in spite of their doubts[88] because they feared being reprimanded for *inertia*, then there is no reason to assume that they were less fearful of being reprimanded for *implementing a solution that would fail*. Having doubts is the same as being aware of the risks of a possible undesired outcome. But by acting as they did, i.e., by not linking their decisions to any third-party expertise and disregarding calls to caution even from Lysenko, the officials literally asked to be identified as scapegoats if failure would occur—which is, obviously, exactly the opposite of what fear and risk-awareness would call for. Thus, (1) can be excluded.

Was the NKZS under significant pressure to find a solution to Black Frost? Available materials suggest that it was not so. First, to the best of my knowledge, none of the historiographical accounts that presuppose administrative pressure quote any governmental decree or order that the NKZS must find a solution to Black Frost.[89] Second, it is not clear why in the year the famous Lenin All-Union Academy of Agricultural Sciences [VASKhNIL][90] was created on the institutional basis of the VIPBiNK, a republican ministry of agriculture rather than the VASKhNIL, the VIPBiNK, or any other institute of plant sciences, would have been tasked to solve a problem that required significant amounts of research. Third, in 1929, Vavilov, who had publicly stated in summer 1928 that the Black Frost problem was solvable only through extensive research, was made head of the VASKhNIL.[91] It is difficult to explain why party-state decision-makers would distrust Vavilov so much as to demand from the NKZS an immediate solution to it and then make him president of the most important agricultural research organization of the Soviet Union. Fourth, the goal of the Ukrainian State Commission created in the summer of 1928 with Shlikhter as its head was to eliminate the *results* of the Black Frost problem, not its *causes*. NKZS officials had clear tasks to fulfill, but these tasks were administrative, not scientific.[92] Fifth, the party-state's short-term solution to the problems with grain procurement was collectivization, not scientific methods to increase grain yields.

If the NKZS, however, was not pressured, why did Shlikhter refrain from conferring with scientific experts? Without additional archival findings, it is impossible to make definite statements. Biographical data can give a possible explanation. As a matter of a fact, the lack of a higher education never prevented him from writing numerous "scientific-propagandistic" (*nauchno-propagandistskie*) works on economics, history, agriculture, Marxist theory, and the agrarian question[93]. His works were all written to demonstrate that the Marxist program and worldview were superior. In 1924, he was made director of the Kharkov Communist University,[94] where he started giving lectures. In 1928, he was elected to full membership of the Ukrainian Academy of Sciences. This election was rigged. Its purpose was to strengthen political control over the academy by making communist candidates—regardless of their often doubtable scientific merits—get elected. Shlikhter was such a political candidate.[95] I believe it very well possible that if a person is self-opinionated enough to write on various scientific subjects and teach at a university in spite of not having a higher education, she might delude herself into thinking that albeit being a "political" candidate, she merited her membership to the Academy in scientific terms. Scientific ignorance, together with an unwarranted, overinflated self-esteem, could amount to a possible explanation of why Shlikhter refrained from cooperation with scientific institutes and made definite claims in terms of the agricultural value of vernalization at a point of time that was too early even in Lysenko's opinion.[96]

The press' influence on the *results* of the evaluation was obviously negligible. Journalists simply quoted NKZS officials. Their unsuccessful attempts to coherently reproduce information they received from the NKZS reflected the wider problem of the epidemic levels of professional incompetence in the pre-war Soviet press: semi-literate journalists and editorial staff were the rule rather than the exception.[97] This problem was exacerbated by the fact that *Pravda*, in particular, was read predominantly by ambitious party members willing to learn the Bolshevik lingo, but lacking basic scientific literacy.[98]

The opinion of the Soviet plant science community also had no significant impact on the NKZS. In 1929, "bourgeois" scientists were not in the best position to denounce the actions of the NKZS and belittle a scientist who was not only repeatedly praised by a state ministry but also had a peasant background. Their position must have motivated them to soften their criticism of Lysenko or, at least, embellish it, as, for instance, Maksimov seems to have done, with incoherent statements of approval or even praise.[99]

As for Lysenko, he had clearly no initial plan to solve the problem of Black Frost and to get famous for doing so. It was the NKZS officials who framed the results of his father's field experiment as a discovery. In the July interview with the *Komunist*, Lysenko took credit only for his theoretical concept of winterness and was clearly less interested in practical questions. During the September meeting, Talanov apparently was under the impression that Lysenko was being unusually modest. This modesty and Lysenko's demonstration of at least some degree of skepticism toward the practical aspects of his work sharply contrast with his arrogant December article. Between September and December 1929, Lysenko must have gone through some significant change. An easy explanation for this change would be that after being promoted to Odessa in October or November, he became fully aware of the amount of institutional backup the NKZS was ready to offer him. If Shlikhter's enthusiasm had already provided him with a senior position at the UGSI, why not see how far he could go?

Unconditional support from the NKZS would explain why by December, Lysenko had ceased to speak of additional tests and begun to attribute explanatory power to his father's field experiment. However, the fact that he also argued that a crucial feature of his method was its relative infeasibility for agricultural use does not fit with him manipulating the NKZS: after all, a conscious attempt to benefit from the influence of a protector is usually made by demonstrating agreement. I believe that Lysenko's playing down of the practical use of his method stemmed from the fact that in November, none of the scientists had described the *theoretical sides* of his work as important, with Maximov stating flatly that in theory, cold germination was the same as Lysenko's theory of winterness. As Roll-Hansen has argued, in his early days, Lysenko was not the cynical opportunist he became later. Rather, he was an ardent believer in his own theories, overly zealous to make a major theoretical discovery and unable to deal with scientific criticisms.[100] As he only began using the term "discovery" in December, he must have been aware that the NKZS was unable to put his work into its correct scientific context. That the ministry neglected his own calls for caution was with no doubt very instructive.

Conclusions

The NKZS was by far the most important promoter of Lysenko. Neither Lysenko nor the media pushed the ministry into embracing his work uncritically. The sources I have analyzed in this chapter strongly suggest that flawed decision-making at the NKZS stemmed from scientific incompetence, rather than pressure from the party-state. To make definitive statements on this question without additional material from the relevant Ukrainian archives is impossible. Further research should focus on the decision-making processes and structures within the NKZS of the late 1920s.

To this end, finding documents revealing how the scientific aspects of Lysenko's work were discussed internally by the NKZS is crucial. Two issues seem particularly important: One, how was the news of the discovery handled initially in mid-July 1929? Two, did someone suggest asking a scientific institute for assistance in the evaluation process, and, if so, how was this overruled? It is also crucial to find documents showing how the NKZS formally adopted decisions.[101] If further archival findings provide conclusive evidence of my view that the NKZS was ill-informed rather than pressured, Lysenko's father becomes a decisive figure in his son's rise to the heights of Stalin's Soviet Union. By informing the NKZS, rather than a scientific institute about the result of his field experiment, he handed his son's seed treatment to a group of people unqualified to assess it. As pressure from the Moscow party elite as a significant factor for Lysenko's early success becomes less likely, individuals and organizations that have been largely ignored become more important. For instance, too little is known about the interactions of the NKZS with the UGSI during the second half of 1929. In this regard, finding archival material which would shed light on Sapegin's contact with the ministry seems particularly promising.[102]

The lesson to be drawn from the waste of Soviet agricultural resources and manpower as a result of the flawed decision-making at the NKZS is not so much about the professional competences state officials should have. Rather, it is about the administrative-political competences they should not have. By nature, bad decisions by state officials are never subject to the clear and unavoidable feedback that bad decisions made in the marketplace usually trigger. Had

Soviet peasants been free from state coercion and worked for their own profit, vernalization would, without any doubt, not have been adopted at all or been disposed of much faster. Because every state administration has its share of incompetent and self-opinionated officials, perhaps we should be more alert about the possible "side-effects" of governments having, for instance, the authority to subsidize renewable forms of energy, or central banks having the power to create money out of thin air. Surely, reducing pollution and stimulating the economy are noble goals. But so was raising Soviet crop yields.

Notes

1. A. Shlikhter, "O poseve ozimykh kul'tur vesnoĭ: (Otkrytie agronoma T. D. Lysenko)," *Pravda*, October 8, 1929, 2.
2. In Ukrainian: *Narodnyĭ komisariat zemel' nikh sprav USRR*.
3. See Shlikhter, "O poseve ozimykh," 2.
4. As it will become clear below, the notion that Lysenko made a discovery is erroneous.
5. Shlikhter referred to any of those three "sub-discoveries" indiscriminately as "Lysenko's discovery" and failed to explain why. I will discuss the probable reasons for this failure in further detail below.
6. See Richard Lorenz, "Die Stagnation der sowjetischen Getreidewirtschaft zwischen 1927 und 1929," *Jahrbücher für Geschichte Osteuropas*, Neue Folge 18 (1970): 389–425, here 408f.
7. See Mark B. Tauger, "Grain Crisis or Famine? The Ukrainian State Commission for Aid to Crop-Failure Victims and the Ukrainian Famine of 1928–29," in *Provincial Landscapes: Local Dimensions of Soviet Power, 1917–1953*, ed. Donald J. Raleigh (Pittsburgh: University of Pittsburgh Press, 2001), 146–170, here 147–162.
8. See David Joravsky, *The Lysenko Affair* (Cambridge: Harvard University Press, 1970), 61.
9. The Bolshevik leaders needed to export large quantities of grain in order to finance Soviet industrialization. During the 1920s, they had run into serious difficulties with peasants not wanting to sell enough grain at the (very low) state-set price. Collectivization—that is, turning the peasants' land into state property and the peasantry into a kind of state labor force—was regarded as the remedy by Stalin.
10. See Markus Wehner, *Bauernpolitik im proletarischen Staat: Die Bauernfrage als zentrales Problem der sowjetischen Innenpolitik, 1921–1928* (Köln: Böhlau, 1998), 94–105, 309–390.

11. See Robert Conquest, *The Harvest of Sorrow: Soviet Collectivization and the Terror-Famine* (New York: Oxford University Press, 1986), 87–188; Andrea Graziosi, "The Great Soviet Peasant War: Bolsheviks and Peasants, 1917–1933," in *Stalinism, Collectivization and the Great Famine*, ed. Andrea Graziosi (Cambridge: Ukrainian Studies Fund, 2009), 5–64, here 33–40.
12. In Russian: *Sel' skokhoziaĭstvennaia gazeta*. In 1929, it was the leading Soviet agricultural newspaper.
13. See NARKOMZEM UKRAINY, "Otvet kritikam: Ob opytakh tov. Lysenko po iarovizatsii ozimykh," *SG*, December 19, 1929, 3.
14. The term "vernalization" (in Russian: "*iarovizatsiia*") appeared only in October 1929; it can refer to a tool for plant-breeding and to an agricultural technique to raise crop yields. For plant-breeding, vernalization has its merits. The application for plant-breeding, however, had virtually no impact on Lysenko's success in 1929; I use "vernalization" exclusively in reference to the agricultural technique. Before October, vernalization was called "Lysenko's method" or "Lysenko's discovery"; after October, all three terms were used interchangeably.
15. In Ukrainian: *Ukraïns' kiĭ genetichno-selekciĭnyĭ instytut.*
16. See Joravsky, *Lysenko Affair*, 61.
17. See Valeriĭ Soĭfer, *Vlast' i nauka: Istoriia razgroma genetiki v SSSR* (Tenafly: Ėrmitazh, 1989), 41. Lysenko later stated that it was in November that the NKZS had invited him to work in Odessa. According to other sources, he was invited in October. In any case, to the best of my knowledge, none of the historical accounts on Lysenko's early years contain any specifications as to the type, title, or date of Shlikhter's order.
18. See Nils Roll-Hansen, *The Lysenko Effect: The Politics of Science* (Amherst: Humanity Books, 2005), 120–154, especially 124–126. During the chaotic early 1930s, field trials on the effectivity of vernalization were conducted under Lysenko's supervision. This set-up gave Lysenko ample room to conceal the reality of the test results. Sometime around 1938, the use of vernalization was discontinued "quietly." Among the party-state elite and (thus) the Soviet public, Lysenko's scientific reputation took no harm.
19. See, for instance, Joravsky, *Lysenko Affair*, ix.
20. Joravsky, for instance, argues that due to the pressure the NKZS was subjected to from "above," one actually could have expected it to

react the way it did (see *Lysenko Affair*, 60f.). Soyfer (*Vlast' i nauka*, 33–41) puts forward a very similar argument. Other accounts do not explicitly state that the NKZS was more or less forced to react the way it did, but, nevertheless, presume a strong connection between the ministry's decision-making and the agricultural problems it supposedly had to solve. See, for instance, Zhores Medvedev, *Vzlet i padenie Lysenko: Istoriia biologicheskoĭ diskussii v SSSR (1929–1966)* (Moskva: Kniga, [1967] 1993), 27; Nikolai Krementsov, *International Science between the World Wars: The Case of Genetics* (London: Routledge, 2005), 83.

21. See, for instance, Joravsky, *Lysenko Affair*, 35–38.
22. See Roll-Hansen, *Lysenko Effect*, 133–192.
23. See Roll-Hansen, *Lysenko Effect*, 119f. Roll-Hansen makes a broader case on how flawed Marxist science politics fostered confusion over the evaluation criteria for basic and applied science in prewar Soviet science. He argues that this confusion repeatedly played into Lysenko's hand. However, he does not explicitly link the 1929 evaluation of vernalization to his broader argument.
24. These sources consist of two articles that were published in the Ukrainian newspaper *Komunist* in late July and the *Bulletin of the NKZS*.
25. D. A. Dolgushin and T. D. Lysenko, "K voprosu o sushchnosti ozimi,"in *Trudy Vsesoiuznogo s"ezda po genetike, selektsii, semenovodstvu i plemennomu zhivotnovodstvu, v Leningrade 10–16 ianvaria 1929 g.* (Leningrad: [?], 1929), vol. 3, 188–199. As I could not consult this publication, I rely on Roll-Hansen's description (see *Lysenko Effect*, 66–68, 72f.).
26. In 1927, a journalist working for *Pravda* visited Lysenko's place of work in Gandzha and decided to write a feature about him. This feature made Vavilov aware of Lysenko and his work. Vavilov sent an employee to Gandzha to investigate on Lysenko. Upon hearing this employee's report, Vavilov wanted to invite Lysenko to his institute. However, Maksimov was against such an invitation, and Vavilov apparently did not see in Lysenko a talent promising enough to upset Maksimov: Lysenko received no invitation. See Semen Reznik, *Nikolaĭ Vavilov* (Moskva: Molodaia gvardiia, 1968), 263–266.
27. In Russian: *Vsesoiuznyĭ institut prikladnoĭ botaniki i novykh kul'tur*. The VIPBiNK was the largest and most important institute of Soviet plant sciences.

28. See Reznik, *Nikolaĭ Vavilov*, 266.
29. See R. O. Whyte, "History of Research in Vernalization," in *Vernalization and Photoperiodism*, ed. A. E. Mureneek et al. (Waltham: Chronica Botanica Company, 1948), 1–38, here 3–5.
30. T. D. Lysenko, *Vliianie termicheskogo faktora na prodolzhitel'nost' faz razvitiia rasteniĭ, opyt so zlakami i khlopchatnikom* (Baku: Trudy Azerbaĭdzhanskoĭ Tsentral'noĭ Opytno-Selektsionnoĭ Stantsii im. tov. Ordzhonikidze v Gandzhe, 1928). I rely on Nils Roll-Hansen's evaluation (see *Lysenko Effect*, 58–64) of this publication as I could not consult it myself.
31. See Roll-Hansen, *Lysenko Effect*, 28–32, 58–64.
32. I rely on Roll-Hansen's summary (see *Lysenko Effect*, 64–68, 72f.) of the congress and of the paper Lysenko presented, as I could not consult it myself.
33. Roll-Hansen, *Lysenko Effect*, 73.
34. This claim Lysenko had made already in his 1928 monograph. See Roll-Hansen, *Lysenko Effect*, 60f.
35. See Roll-Hansen, *Lysenko Effect*, 64–68, 72f.
36. Soyfer (*Vlast' i nauka*, 35) doubts the account of Lysenko Sr. contacting the NKZS, arguing that the distance between Karlovka (where his farm was) and Kharkov would be too big for the story to be true. However, Shlikhter did not state that Lysenko Sr. went to Kharkov: he probably contacted a nearer, regional branch.
37. How Lysenko, who presumably was working in Gandzha, was notified of the events and when he arrived in Ukraine is not clarified in any of the accounts I know of.
38. See Shlikhter, "O poseve ozimykh," 2. Shlikhter's timeline of these mid-July events does contradict one other account, but it coincides with yet others and it does not contradict verifiable facts that I know of. However, Shlikhter's article contains a timeline of another event that does contradict actual facts. Therefore, Shlikhter's description of the mid-July events should be regarded as likely true, but not as face value.
39. See Vl. Grigor'ev, "Otkrytie agronoma Lysenko: Metod Lysenko budet na praktike primenen v sovkhozakh i kolkhozakh Ukrainy," *Pravda*, July 21, 1929, 4.
40. I was unable to find biographical information on Gorban'.
41. The story contained very unlikely to-have-happened elements which in subsequent accounts of the story were never mentioned again.

42. In Russian: "*Perspektivy* [...] *nastol'ko veliki, chto ne poddaiutsia srazu skol'ko-nibud' deĭstvitel'nomu podschetu.*"
43. In Russian: "*Urozhaĭ iarovoĭ 'ukrainki,' po opredeleniiu komissii, dolzhen vyrazit'sia priblizitel'no okolo 3 tonn s gektara.*"
44. Grigor'ev, "Otkrytie Lysenko," 4.
45. The *Komunist* was the organ of the Ukrainian Communist Party's central committee; it was edited in Kharkov.
46. In December 1928, Lysenko gave a report on the influence of low temperatures on development stages of plants in Kiev (see Soyfer, *Vlast' i nauka*, 31). Lysenko probably visited his father on his trip from Gandzha to Kiev or from Kiev to Leningrad.
47. See "Praktychni doslidy bat'ka tov. Lysenka," *Komunist*, July 28, 1929, 2.
48. "Tsinnyĭ vinakhid u haluzi sil's'koho hospodarstva: Chy mozhna siiaty ozyminu vesnoiu?," *Komunist*, July 28, 1929, 2.
49. Italics are added.
50. See "Tsinnyĭ vinakhid...," 2.
51. "Tsinnyĭ vinakhid...," 2.
52. In Russian: *Ėkonomicheskaia zhizn'*. The *ĖZh* was a newspaper of the NĖP-era, mainly read by the nėpmeny and state factory managers seeking financial and trade information.
53. "Otkrytie agronoma Lysenko: V chem sut' otkrytiia," *ĖZh*, August 4, 1929, 4.
54. See Kachinskiĭ, "Strakhovka ozimykh ot vymorazhivaniia," *ĖZh*, August 4, 1929, 4.
55. While the term "*zamorazhivanie*" can denote "chilling," it is used mainly for processes that cool a substance to temperatures below the freezing point of water. Remember that Lysenko had used the correct term "*okholodzhennia*" (in Russian: "*okhlazhdenie*") a week earlier, when he described his experiments to the reporter of the *Komunist*.
56. Of course, the prefixes 'vy-' and 'za-' connote slightly different things. However, the respective meanings of the two words are so close to each other that in the given context, it is difficult to understand why anyone literate to at least some extent—and even without having a clear understanding of the seed treatment in question—would not see the obvious: that the problem and the solution to the problem are more or less the same thing.
57. Mashura, "Nuzhny dal'neĭshie izyskaniia," *ĖZh*, August 4, 1929, 4.

58. Filipovskiĭ, "Chto dolzhen delat' Sakharotrest," *ĖZh*, August 4, 1929, 4.
59. I was unable to find dates of birth and death.
60. TsGANTD, Sankt Peterburg, f. 318, op. 1, d. 230, l. 95. Vavilov was not present. In fact, he had no influence whatsoever on the promotion of Lysenko's "discovery," as he was traveling in Eastern Asia from June to December. See Semen Reznik, "Snova o Vavilove i Lysenko," *Priroda*, no. 11 (1992): 88–98, here 91.
61. From the protocol, it is clear that the matter in question was the experiment on the farm of Lysenko's father. The "8" in "1928" certainly is a misprint.
62. See TsGANTD, Sankt Peterburg, f. 318, op. 1, d. 230, l. 96.
63. Remember that Maksimov had pointed this out already in the aftermath of the January congress.
64. Lysenko stated that due to urgent work, he could not write such a report right away; nevertheless, he hoped that he could publish one soon. Why such a report never was published, I cannot say.
65. See Dmytryĭ Ursu, "Genetika v Odesse: sto let bor'by, pobed i porazheniĭ," *Iugo-Zapad. Odessika* 14 (2012), 210–257, accessed July 6, 2014, http://dspace.onu.edu.ua:8080/bitstream/123456789/2589/1/Генетика%20в%20Одессе-прав.%20Урсу%2B.pdf/. The September 1 meeting is quoted in quite a few accounts of Lysenko's early years. However, to the best of my knowledge, none of these accounts mention any follow-up action on the part of the VIPBiNK. As, however, none of these accounts include a statement to the effect that no such follow-up actions have taken place, I believe that further research at the TsGANTD (which I was not able to undertake) might reveal valuable information on what happened between the September 1 invitation of Lysenko to work at the VIPBiNK and his transfer to Odessa in October.
66. See Shlikhter, "O poseve ozimykh," 2.
67. L. Emel'ianenko, "Iarovizatsiia ozimykh kul'tur: Po povodu opytov T. D. Lysenko," *SG*, October 10, 1929, 3.
68. Emel'ianenko stated that he was part of the commission that inspected the field of Denis Lysenko. However, the wording of this statement was such that it did not follow from it that Emel'ianenko was affiliated with the NKZS.
69. Note that the 1929 issues of the *Zbirnyk zakoniv ta rozporiadzhen' robitnycho-selians'koho uriadu Ukraïny*, in which official decisions of

the Ukrainian Council of Ministers were published, contains no reference to Lysenko.

70. See *Biuleten' Narodn'oho Komisariiatu Zemel'nykh Sprav*, November 8–15, 1929 (Issue 42, No. 181), "Obizhnyk NKZS za 3/XI-29 r. No 50/14-0/561 'pro poriadok perevedennia doslidchykh posiviv na selians'kykh zemliakh za sposobom ahronoma T. D. Lysenka'," Moscow State Library, Khimki Branch.
71. P. Lisitsyn, M. Prik, N. M. Tulaĭkov and A. A. Sapegin, "Iarovizatsiia ozimykh," *SG*, November 13, 1929.
72. Remember that Kachinskiĭ spoke of *zamorazhivanie*, which, as *promorazhivanie*, stems from *moroz* (frost).
73. See N. Maksimov, "Vesenniĭ posev ozimykh otkryvaet shirokie perspektivy," *SG*, November 19, 1929, 3.
74. Batyrenko did not declare that on September 1, it was Lysenko who said that the results of his father's experiment must probably be attributed to lucky circumstances.
75. See V. T. Batyrenko, "Shiroko razvernut' proverochnye opyty po metodu Lysenko," *SG*, November 19, 1929, 3. The "T." (instead of the correct "G.") is a misprint.
76. "Iarovizatsiia ozimi–novoe zavoevanie v bor'be za urozhaĭ," *SG*, November 19, 1929, 3.
77. The way Lysenko used the word "hypothesis" suggests that he thought it to be a synonym for theory.
78. Lysenko, "V chem sushchnost' gipotezy 'ozimosti' rasteniĭ," *SG*, December 7, 1929, 3.
79. Lysenko referred to both his seed treatment and his theoretical speculations on plant development as "hypothesis of winterness," as "method of winterness" and as "method of vernalization." He failed to properly define any of the three terms. Also, his comparison of his theory to cold germination reveals a certain degree of confusion over what Maksimov had written in his article.
80. Lysenko, "V chem sushchnost'," 3.
81. See Lysenko, "V chem sushchnost'," 3.
82. NARKOMZEM UKRAINY, "Otvet kritikam," 3.
83. For details on the "politicized" criticisms Soviet scientific literature was permeated with by the end of the 1920s, see Nikolai Krementsov, *Stalinist Science* (Princeton: Princeton University Press, 1997), 24–29.
84. No possible reason or motive for such a conspiracy was given.

85. One such attempt to appear knowledgeable was "V samom dele, i po uchebniku Prianishnikova nam izvestno, chto est' sposob inogda poluchit' iarovizirovannye ozimye, no pravil'no otmechaet akad. Sapegin, chto iarovizirovat' ozimye udavalos' ne vsegda." Unfortunately, I am not able give a translation that would convey how self-defeating this attempt was. Suffice to say that the greater part of the article could have been penned by a short story figure of Mikhail Zoshchenko.
86. See D. F. Virnyk, "Zhizn' i revoliutsionnaia deiatel'nost' A. G. Shlikhtera," in A. G. Shlikhter, *Agrarnyĭ vopros i prodovol'stvennaia politika v pervye gody sovetskoĭ vlasti* (Moskva: Nauka, 1975), 20–38, here 21 f.
87. Remember that Emel'ianenko mentioned earlier research (by, among others, Maksimov) before the *SG* printed the scientists' opinions, and that he was part of the commission that inspected Denis Lysenko's field. Judging by his overall literate and informative October account, I believe that he was hardly under the impression that the field experiment was exiting news. Obviously, for reasons of hierarchy or others, his knowledge was not taken into due consideration by his superiors.
88. Of course, the extent of the officials' scientific ignorance makes this scenario very unlikely.
89. This is counter-intuitive insofar as the party-state did indeed assign (scientific) organizations with tasks of agricultural sciences by formal decrees. See, for instance, Prezidium TsKK VKP(b) and Kollegiia NK RKI SSSR, "O selektsii i semenovodstve," *Pravda*, August 3, 1931, 3.
90. In Russian: *Vsesoiuznaia akademiia sel'skokhoziaĭstvennykh nauk imeni Lenina*.
91. See Roll-Hansen, *Lysenko Effect*, 89f.
92. See Tauger, "Grain Crisis," 152.
93. The communist universities were set up as the Bolshevik alternative to "bourgeois" science. They were organized in the hierarchical way of the Communist Party; on all of these universities' matters, party cells decided. See Krementsov, *Stalinist Science*, 24f.
94. The agrarian question comprised, first and foremost, socioeconomic, not technical, aspects of agricultural production and peasant life.
95. See L. V. Matveeva and Ė. G. Tsygankova, "Vseukrainskaia Akademiia nauk. God 1929-ĭ," in *IN MEMORIAM: Istoricheskiĭ sbornik*

pamiati F. F. Perchenka, ed. A. I. Dobkin et al. (Moskva: Feniks, 1995), 112–140, here 121–127; A. M. Rumiantsev and V. V. Oreshkin, "Vydaiushchiĭsia revoliutsioner i uchenyĭ-ėkonomist," in: A. G. Shlikhter, *Agrarnyĭ vopros i prodovol'stvennaia politika v pervye gody sovetskoĭ vlasti* (Moskva: Nauka, 1975), 7–19; Virnyk, "Zhizn' Shlikhtera," 35–37.

96. Maybe, Shlikhter's line of thought was similar to the following: Lysenko was an agronomist. He, Shlikhter, was a full member of the Academy of Sciences. If Lysenko said that further tests were necessary, then he was right. In science, there are always tests that need to be done. So someone must do them. He might not have the time or specific knowledge to organize them, but that was why he had specialists such as Emel'ianenko. However, why wait around for results confirming the obvious? The facts were on the table. The wheat on Lysenko's farm did not grow so well by chance, but because of the discovery. That is why it must be implemented as fast as possible. To struggle with some hair-splitting bourgeois experts—who had failed to understand Lysenko's work properly already at the January congress and were not even *akademiki*—was pointless.
97. See Julie K. Mueller, "Staffing Newspapers and Training Journalists in Early Soviet Russia," *Social History* 31 (1998): 851–873.
98. See Jeffrey Brooks, *Thank you, Comrade Stalin! Soviet Public Culture from Revolution to Cold War* (Princeton: University Press, 2000), 5, 12f, 17; Alexei B. Kojevnikov, *Stalin's Great Science: The Times and Adventures of Soviet Physicists* (London: Imperial College Press, 2004), 280. For an account of how the incoherent *Pravda* coverage of the infamous IV session of VASKhNIL in December 1936 played into Lysenko's hands, see Lukas Joos, "Der Aufstieg T. D. Lysenkos zum Präsidenten der VASChNIL im Spiegel der ‚Pravda'-Berichterstattung: Zur Darstellung von wissenschaftlichen Inhalten im Organ des Zentralkomitees der KPdSU(B) in den Jahren 1927–1938" (Lizentiatsarbeit, University of Zurich, 2011), 110–138.
99. Why none of the VIPBiNK scientist questioned Lysenko on his theoretical conclusions during the September 1 meeting is something I do not know cannot answer. Having Tulaĭkov's and Lisitsyn's sharp public criticisms in mind, to contend simply that the entire group of scientists was too afraid to ask Lysenko critical questions seems wrong to me. However, as the meeting's minutes do not include a

complete list of participants, it cannot be excluded that someone attended whom the VIPBiNK scientists distrusted.

100. See Roll-Hansen, *Lysenko Effect*, 56, 69.
101. For example, Shlikhter's decree on Lysenko's transfer to Odessa and a possible formal adoption of the July decision to use Lysenko's discovery on Ukrainian state farms.
102. There is no reason to believe that in other key events of Lysenko's rise, professional incompetence of the party-state administrators did not play a significant role. For the IV session of VASKhNIL of December 1936, Roll-Hansen provides a quite telling source which points exactly in this direction. He quotes a letter from the American geneticist Hermann J. Muller (1890–1967) who participated in the session. In this letter, Muller described how, in the aftermath of the session, the minister of agriculture and the head of the science branch of the party's central committee engaged him in an discussion over "gene constancy" that revealed both the officials' stunning incompetence and their unwarranted self-regard. However, Roll-Hansen does not elaborate on the implications of top party cadres simply not knowing what they were talking about. See Letter from Muller to Huxley, March 9, 1937, Muller archive, cit. after Roll-Hansen, *Lysenko Effect*, 214.

Pavel Pantelimonovich Luk'ianenko and the Origins of the Soviet Green Revolution

Mark B. Tauger

The idea of a "Soviet Green Revolution" seems almost inconceivable: according to almost all scholarly and popular publications, the Stalinist policy of collectivization devastated Soviet agriculture, and the pseudo-scientist Trofim Lysenko, supported by Stalin and Khrushchev, similarly destroyed Soviet genetics. Recent studies, however, have challenged this stereotype about collectivization.[1] This article deals with Soviet genetics and plant breeding by describing and contextualizing the hybrid wheat-breeding work of Soviet scientist Pavel Panteleimonovich Luk'ianenko (1901–1973) at the Krasnodar agricultural research institute in the Kuban region. Luk'ianenko and his associates created wheat varieties resistant to several of the environmental threats that had long caused crop failures and famines in Kuban and elsewhere in the USSR. These varieties included

This paper was originally presented at the Second International Workshop on Lysenkoism, 22–23 June 2012, University of Vienna, Vienna, Austria, which was funded in part by the US National Science Foundation. My participation also received support from West Virginia University. I also would like to thank Professor Joginder Nath of the Davis College of Agriculture at West Virginia University for very helpful discussions.

M.B. Tauger (✉)
West Virginia University, Morgantown, WV, USA

W. deJong-Lambert, N. Krementsov (eds.),
The Lysenko Controversy as a Global Phenomenon, Volume 1,
DOI 10.1007/978-3-319-39176-2_4

semi-dwarf high-yielding varieties [HYV] of wheat—such as Bezostaia-1—which achieved international recognition for their high quality.

Luk'ianenko's work, which continued the work of the great Soviet biologist Nikolai Vavilov, was clearly on the forefront of world applied genetics, yet it took place in the USSR during the peak period of Lysenko's power. This Soviet breeding program began before the plant breeding work of Norman Borlaug in northwest Mexico that was later called the "Green Revolution," and continued contemporaneously with it. The Soviet program had the same objectives, employed the same methods based on standard genetic principles, and used many of the same sources of plant germplasm, including hybrid dwarf and semi-dwarf wheats from Japan, Italy, Argentina, and the USSR. The concurrence of Lysenko's dominance and Luk'ianenko's initiation of a Green Revolution in the USSR, absent from the historical literature except for a single brief reference in one publication, raises important questions about the conventional interpretation of Lysenkoism and Soviet agriculture.[2]

This chapter outlines the main components of the Green Revolution outside the USSR and describes and analyzes the main Russian and Soviet research that laid the basis for similar developments in the USSR. It then discusses the limits of Lysenko's attempts to stop Soviet genetics research: It was possible for an innovator to emerge despite Lysenko's administrative position and his efforts to control Soviet research in genetics and agricultural sciences. Finally, the chapter then discusses Luk'ianenko's work and important issues that this history raises.

The International Green Revolution

The history of the Green Revolution commonly known in America focuses on plant scientists from the USA.[3] To summarize this history very concisely, Mexican officials asked the Rockefeller Foundation for aid in 1942, after infestations of rust, a major fungal plant disease of wheat, devastated wheat crops, and greatly reduced harvests three years in a row. The Foundation financed a group of scientists who began a research program in Mexico to breed wheat resistant to disease and drought. One of these scientists, Norman Borlaug (1914–2009), set up an experiment station in Ciudad Obregón in northwest Mexico, and in the 1950s, brought into its breeding program the partly Japanese dwarf wheat Norin 10. By the early 1960s, Borlaug's group had bred semidwarf high-yielding varieties (HYV) that produced much higher yields when used with a "package" of fertilizer, pesticides, and irrigation. These varieties inherited from Japanese wheat one or more "reduced height genes" (Rht), which caused the plants to grow short but very thick and sturdy straw that would be much less likely to "lodge" or collapse.[4] This crucial phenotype transformed these plants into HYV because it allowed farmers to use more fertilizer so that the plants could develop a larger ear of grain containing more and heavier seed without the risk that lodging would destroy the crops. In 1963, the Ciudad Obregón center was renamed the International Center for the Improvement of Maize and Wheat, known by its Spanish acronym, CIMMYT.

Borlaug encouraged international exchanges of varieties to develop new ones and proselytized among the world leaders to adopt the HYV. After two years of droughts, crop failures, and incipient famine in 1965–1966 in India, Borlaug persuaded Indian politicians to import HYVs and expand breeding programs. By 1970, India dramatically increased wheat production. For this success as well as his previous work, Borlaug was awarded the Nobel Peace Prize in 1970. In 1971, CIMMYT and other recently formed research centers were unified under the Consultative Group on International Agricultural Research (CGIAR), which now includes 16 agencies working on a wide range of development problems.

This American-centered narrative of the Green Revolution omits or minimizes plant breeding work done before Borlaug and outside the USA in an international effort to produce higher-yielding crops. A brief review of this earlier history provides the necessary background to Luk'ianenko's work and helps understand his unique contribution.[5] Japanese farmers obtained the short wheat varieties in the sixteenth century from Korea, where they had been grown since late antiquity.[6] In the 1920s, Japanese breeders crossed a short variety with varieties from the Mediterranean and Russia to produce 18-inch-high Norin-10.[7] Meanwhile, the first European to incorporate Japanese varieties and breed what were, in essence, the first HYVs was the Italian scientist Nazareno Strampelli (1866–1942).[8] Strampelli set out to breed high-yielding wheat because of Italy's dependence on imported wheat, as in 1904, when crop failures caused shortages in Italy while Argentina, its main source of imports, had an enormous harvest. Strampelli worked to make Italy self-sufficient in wheat years before Mussolini came to power and began his "Battle for Wheat" campaign of the 1920s.[9]

Strampelli and his coworkers at his research center in Rieti produced at least 65 new wheat varieties that substantially increased wheat production in Italy between 1900 and World War II. He specifically set out to breed wheat that had short stems to resist lodging, ripened early, and resisted rust, using the Japanese dwarf wheat Akakomugi. Two important results of these crossings were the varieties Ardito and Mentana, released in the 1920s and used on millions of acres in Italy and elsewhere. In 1923, Strampelli went to Argentina with some of his varieties and inspired agronomists there to breed varieties based on the Italian ones but suited to their conditions. One breeder used Ardito and Mentana in crossings that produced rust-resistant semi-dwarf varieties, named Klein 30, 31, and so forth, after the scientist who developed them. Borlaug used Mentana as one of his main breeding varieties in Mexico in the 1940s and applied the same basic approaches as Strampelli. The work of Strampelli and the other Italian breeders was (as an Italian scholar put it) the "first example" of the Green Revolution.[10]

As the examples of Italy and Mexico show, the Green Revolution was inspired by chronic low food production, vulnerability to crop failures, and dependence on imported food. The efforts to develop new varieties involved cross-breeding of diverse domestic and foreign varieties to create new types of wheat with short, sturdy stems to resist lodging and with resistance to plant diseases, severe weather, and other characteristics. These characteristics were necessary to enable these new varieties to produce much higher yields.

The Russian Background to the Green Revolution and the Work of N.I. Vavilov

The history of plant breeding in Russia and the USSR followed more or less the same pattern as the Green Revolution in the west: it sought the same goals, used many of the same varieties, and anticipated Borlaug's work. Like Mexico and Italy, Russian breeding was also motivated by agricultural crises. The main difference was the interruption of Stalinism and Lysenko, discussed in the next section below.

Russia has a long history of famines: famines occurred in more than 400 of its 1000 years of recorded history.[11] Environmental disasters almost always caused these famines: drought or heavy rain, extreme cold or heat, lodging from heavy rains or excessive plant growth, weeds, pests, and blights (plant diseases). Russians, from peasants to scientists, tried to alleviate and prevent these famines through rituals, food imports, elaborate relief systems, and use of resistant varieties of wheat and other crops.

Some varieties were relatively resistant to certain environmental threats, such as the drought-resistant "Turkey Red," as well as other varieties that Mennonite migrants and Mark Carleton (1866–1925) of the USDA brought back to the USA in the nineteenth century.[12] But severe and complex environmental disasters continued to cause crop failures and famines into the twentieth century.

From the eighteenth century, Russian nobles, scientists, and officials established agricultural test plots and ultimately experimental stations, in part to create new varieties of grains and other crops.[13] The number of such plots and the competence of their personnel grew as biological knowledge expanded, and as the regime freed the serfs and founded the zemstvos, rural self-government agencies. Repeated famines, especially the 1891 famine in the Volga region, motivated the government to expand this network of experiment stations and other institutions for agricultural research and education. The trained personnel working in these institutions were educated in Western biological sciences, studied the research done in Western Europe and the USA, and traveled there to observe their farms and methods.

These efforts to create new crop varieties also involved searching for unknown wild and cultivated crop varieties with resistance to environmental threats in Russia and in other countries. The most important Soviet scientist to undertake this phyto-geographical research was Nikolai Vavilov (1887–1943).[14] From early on in his work, including both his dissertation and his first expedition to Persia in 1916, Vavilov repeatedly studied plant infestations and diseases with, the objective of finding resistant varieties.[15] Vavilov's expedition to Persia became the first of six large plant-prospecting expeditions in his career, followed by exploratory trips to the Mediterranean, East Africa, Afghanistan, East Asia, and the Americas. During 1923–1940, Vavilov and his associates went on 180 expeditions, 140 in the USSR, and 40 abroad. By 1940, they had collected more than 250,000 plant varieties and produced several substantial publications on plant resources in different world regions.[16]

The plant and seed collections were stockpiled at a central government research institute dedicated to collecting plant varieties and plant breeding that underwent reorganizations and name changes until it became the All-Union Institute of Plant Production (VIR) in the 1930s. Vavilov headed this agency from 1920 until 1940.[17] These collections, which Vavilov called the "world collection," comprised the first Russian large-scale stockpile of landraces (local native varieties) of many crops and their wild relatives in Russia and abroad, and included hybrid varieties created by plant scientists in many countries. In plant collecting, as in other areas, Russia and the early USSR lagged behind the major European powers and the USA. These efforts by Vavilov and his colleagues at VIR helped the USSR to catch up with Western countries in this area of plant research.[18] Their work also laid the basis for the Soviet Green Revolution because later breeders used many varieties that Vavilov collected to create Soviet HYVs.

In 1920–1921, while Vavilov was establishing VIR, the USSR endured two extensive crop failures from drought and other environmental factors, exacerbated by the disruption of the Soviet civil war. The result was the largest famine in Russian history up to that time. Soviet leader V.I. Lenin responded by accepting aid from the American Relief Administration under Herbert Hoover, as well as by importing food, independently distributing it through a large Soviet-run relief program to both rural and urban populations.[19] The Soviet government, specifically the People's Commissariat of Agriculture of the Russian Republic (NKZ RSFSR) also responded to the famine with a large-scale program to breed and introduce new drought-resistant and higher yielding grain varieties. It began with a decree issued by Lenin in June 1921 that ordered the agricultural agencies to compile a register of crop varieties, accumulate a stockpile of "selected seed"—seed of known origin that reliably produced good yields—and to expand the work and network of experimental stations to produce improved seed and crops.

This famine was the first of three major famines that struck the USSR in the 1920s. The vast famine of 1920–1923 encompassed the Volga valley, Ukraine, and certain other regions. In 1924–1925, a milder crop failure impacted the same regions and caused widespread starvation, and in response, the Soviet government again formed a relief commission, imported food, and fed more than 12 million peasants as well as part of the urban population. In 1928–1929, a major crop failure in Ukraine again required the government to import grain, set up a relief agency, and feed hundreds of thousands of starving peasants. This famine was one of the events that pushed the Stalin regime to implement major changes in agricultural policy, including—specifically—collectivization.[20]

In response to these crises, the Soviet regime intensified its efforts to improve its agriculture scientifically. In 1921, Vavilov went to the USA and worked with the USDA to obtain American seed to help Soviet farmers increase their crops in the wake of the famine.[21] In December 1924, another government decision ordered the agricultural and trade commissariats to set aside about 900,000 tons of selected seed as reserves over the next five to ten years and the complete replacement of all seed used in the USSR with improved varieties. This project turned out to be impossible to achieve in that period. In December 1927, the government prepared a five-year plan for seed production. Finally, in 1929, the regime established a central academy of agricultural sciences, VASKhNiL, headed by Vavilov, to coordinate and advance agricultural research.[22] In its objectives and its management of research institutes in multiple branches of agriculture, though not in its hierarchical and bureaucratic character, VASKhNiL anticipated CGIAR.[23]

The plant-collecting expeditions of Vavilov and his associates in VIR played a central role in expanding Soviet breeding programs. In the expeditions, Vavilov and his associates sought not only new varieties but also clues to the geographical origins of cultivated plants. He suspected that these centers of origin would be the most likely places to find diverse and productive varieties. With such varieties, Vavilov envisaged a vast project to breed crops so productive and resistant to environmental threats that they could end famine in the USSR.[24] The regime also sent Soviet scientists abroad to obtain plant breeding information and collaborate with leading foreign scientists.[25]

Vavilov also published a small book, *The Scientific Bases of Wheat Breeding*, based on the VIR wheat collection.[26] This book surveyed the main wheat varieties grown in the major producing countries, the improved varieties developed in several of these countries, as well as various technical aspects of wheat breeding. To draw all these findings together, Vavilov included a chapter on "The Ideal Wheat Variety" that included a long list of characteristics that an ideal variety would have. These included high grain yield and quality, resistance to drought and other weather threats, stiff straw for resistance to lodging, resistance to plant diseases and awnless heads—because growing awns, the thin stalks protruding from the grain husk, diverted the plant's energy from growing grain.[27]

Soviet agriculture in the 1920s resembled that of Italy in the 1900s and Mexico in the 1940s. The USSR faced environmental crises that caused crop failures and a series of famines, and repeatedly resorted to foreign relief and food imports at considerable expense. The USSR established and expanded institutions for agricultural research, recruited scientists and plant breeders, and promoted international collaboration and exchange of information and varieties. Most important, the Soviet government supported Vavilov's efforts to accumulate a vast collection of plant varieties from around the world to use as a basis for an extensive program of plant breeding. Vavilov's guidelines for "the ideal wheat variety" derived from Western as well as Russian and Soviet breeding goals. Soviet agricultural science thus, by 1930, had both needs and preparations similar to those in Italy and in Mexico, basically the same preconditions for launching a Soviet green revolution.

Stalinism and Lysenko

The Stalin regime began forcibly collectivizing peasant farms in 1929. This was, in part, a highly repressive policy, involving the "dekulakization" (expropriation and exile), and in some cases, imprisonment or execution of peasants whom the regime and its personnel considered to be a threat, on the basis of a Marxist theory of class struggle. This violence and the administrative control that collectivization extended over agriculture led some scholars to view collectivization as a new version of serfdom. Yet in the context of repeated famines, famine relief, and other agricultural policies, collectivization was, first of all, part of a broader program to modernize and industrialize Soviet farming.[28] The regime accompanied collectivization with a large program to accelerate the development and impact of the agricultural sciences, including not only the establishment of VASKhNiL but also of many new experimental stations (1300 existed officially by 1932). The government was thus beginning to implement Vavilov's plans for plant breeding. But Soviet leaders also demanded much faster results than could realistically be expected.[29]

This was the context for the political rise of Trofim Denisovich Lysenko and his followers.[30] The conventional view is that Lysenko was an incompetent scientist or pseudo-scientist, possibly a fraud (in the sense that it is uncertain whether he actually believed what he wrote and said), who caused an unmitigated disaster for Soviet genetics and other branches of biological sciences. First, he and his cronies persecuted competent scientists in many fields of biology on the basis of politicized pseudo-scientific ideas. Many were forced out of their positions, some were imprisoned and ultimately executed, while others were driven to suicide or died in part from the stress of unjustified attacks, interrogations, and threats. Lysenko replaced them with followers who were as incompetent as he was, if not worse.

Next, Lysenko rejected the emerging science of Mendelian genetics, dismissed any research or ideas from European and American scientists as "bourgeois science," and allegedly prevented Soviet scientists from conducting legitimate research in virtually all areas of biological sciences. He used his powerful position in the Soviet science administration to impose simple-minded, incorrect, and often absurd views and approaches in virtually all aspects of biological research and education, until his ouster in 1965. Noël Kingsbury writes in his history of plant breeding, for example, that "The impact of Lysenko on plant breeding and other genetically based technologies was devastating. It has even been suggested that Lysenko's influence so damaged Soviet agriculture that it, more than any single factor, led to the demise of Communism." Kingsbury also claims that Soviet genetics lost three decades because of Lysenko.[31] Valerii Soyfer, in his detailed and substantial study of Lysenkoism, similarly describes Lysenko's impact as devastating.[32]

In this discussion, I do not mean to minimize the suffering and injustices endured by those whom Lysenko and his followers victimized. Nonetheless, several studies have presented substantial evidence that qualifies the impact of Lysenkoism on Soviet scientists and on research in agricultural sciences. In his impact on scientists, while some victims of Lysenko suffered tremendously, other Soviet scientists and agronomists opposed Lysenko openly without suffering significant consequences, and many others managed to evade his sanctions. Joravsky conducted a prosopographical study of Soviet biologists and agricultural scientists and found that only a small percentage of these scientists were subjected to "repression."[33] Most of Lysenko's opponents were not arrested, though some lost their jobs.[34] Many, if not most, of the scientists whom Lysenko had removed from posts found work in scientific fields, and some even continued publishing using various subterfuges.[35] Krementsov argued that scientific and political considerations, such as patronage of powerful individuals, protected many biologists and geneticists from dismissal. They often managed to continue their work in conventional genetics by writing research plans in ways that deceived bureaucratic censors. They could publish their work because journals carried brief articles on Lysenkoist themes as camouflage.[36]

A significant literature has also qualified earlier extreme views regarding Lysenko's impact on the ideas and practices of the agricultural sciences. Lysenko's early ideas were not particularly extreme in the context of biological sciences in the late nineteenth and early twentieth centuries. Roll-Hansen showed that many biologists and agricultural scientists, in the USSR and outside, questioned "neo-Darwinian" conceptions of incremental evolution and suspected that environment did play a role in evolution, even if neo-Lamarckism seemed unacceptable to most of them.[37] Vavilov encountered these disputes as a student in the 1890s–1900s.[38] From the 1930s onward, when Lysenko rejected "Western" biology and advocated extreme anti-scientific views, Joravsky, Krementsov, Roll-Hansen, and others document how several scientists and officials resisted his ideas and his politicization of science. Ethan Pollock notes that even Stalin, in a late publication, condemned the "Arakcheev" [dictatorial] practices in science, implied that Soviet scientists should learn from foreign science, and called for open debate.[39] Stalin's essay clearly challenged Lysenko and gave rise to further challenges. Even while Lysenko was in power, some of his programs failed so visibly that they were openly criticized and reversed, as, for example, his idea of cluster planting trees in the Soviet shelter-belt program of 1949–1952.[40] After Stalin's death, Lysenko and his followers steadily lost power and posts while legitimate geneticists staged a comeback, which implied that many of these scientists and their ideas survived.[41]

Lysenko's dominance thus did not expel all legitimate scientists, stop all legitimate research in Soviet biology, or eliminate opposition to his ideas (much less cause the collapse of the USSR). In order to provide a more specific context for Luk'ianenko's work, the following section presents two concrete examples of these limitations on Lysenko's influence. First, the conflict between Lysenko and Vavilov provides an example of how legitimate scientific ideas could survive such a scientist's death even under Lysenko. Second, Lysenko's pet project of vernalization was not quite as absurd as it seemed.

Vavilov initially supported Lysenko, who rose from the peasantry, obtained a limited agronomic education, and was championed by the Soviet press for his early research. Vavilov could see Lysenko's ignorance but was impressed by Lysenko's energy and commitment. He also thought that Lysenko's artificial vernalization (see below) could be a useful technique for plant breeding.[42] But by 1935, Lysenko turned against Vavilov as an obstacle to his advancement. Lysenko's sycophantic promises to Stalin to breed new varieties rapidly, even when he admitted he needed more time, sounded better to Stalin than Vavilov's more cautious but more honest warnings that progress would take time.[43]

After a series of public disputes between Lysenko and Vavilov and their followers, and conspiratorial politics behind the scenes, the government removed Vavilov from the presidency of VASKhNiL in 1936, and appointed Lysenko to that post in 1938. Lysenko attacked Vavilov publicly and behind his back in meetings with Stalin and other leaders. Vavilov finally defended himself and attacked Lysenko in 1939, but by then, fewer scientists supported Vavilov while increasingly key politicians, including Stalin, supported Lysenko.[44] In 1940, the NKVD unexpectedly arrested Vavilov while he was on an expedition to western Ukraine, newly acquired under the Molotov-Ribbentrop Pact. The NKVD interrogated him, and the Military Collegium of the Soviet Supreme Court sentenced him to execution. Several leading scientists, including the agrochemist Dmitrii Nikolaevich Prianishnikov, appealed to the Soviet leaderhips to spare Vavilov. In 1941, NKDV Commissar Lavrentii Beria decided not to execute Vavilov but allow him to continue work as an imprisoned scientist. When the Nazi invaders reached Moscow, Vavilov and many other prisoners were dispatched to the Volga region, where in January 1943 Vavilov died in prison, ironically of starvation as a consequence of the Nazi invasion.[45]

Yet despite Vavilov's tragic death, Soviet scientists at VIR revived his program of expeditions to find new plant varieties in 1946, despite the Nazi's destruction of some VIR laboratories and fields. From 1946 to 1965, in other words, during the period of Lysenko's dominance and his fall from power, the Institute conducted 130 domestic and (from 1954) foreign expeditions, from 1963, often in collaboration with foreign scientists, and its staff members collected some 200,000 more new plant specimens.[46] In addition, the institute revived in 1952 the practice of geographical plantings, to test varieties from the world collection that Vavilov had begun in 1923 and that survived both the war and Lysenko. VIR also revived Vavilov's program of research and publications on the characteristics and genetics of different categories of plants, with a large book on perennial leguminous grasses in 1950.[47]

This incomplete list of VIR's postwar work indicates that enough of Vavilov's colleagues and students survived the peak of Lysenkoism (at least in part by means of the subterfuges discussed above) to maintain VIR's core components, especially its world plant collection, and restore its scientific role along the lines Vavilov had laid out in his two decades of leadership between the wars.

As for vernalization, Lysenko misunderstood the causes of this widespread growth phenomenon in plants. Most grain plants (as well as many other plants) fall into one of two "habit" categories: spring habit and winter habit. A winter habit plant such as winter wheat or winter barley, planted in fall, needs a period of cold weather within a certain temperature range in order to mature and produce flowers and seed the next spring. "Vernalization" is the term for this process. Lysenko first acquired recognition by reporting on his experiments of wetting and cooling winter wheat plants in early spring to induce them to sprout and then planting the sprouted seed in spring. They would then, in some cases, flower and produce grain. Strictly speaking, this "technique" should be called "artificial vernalization." European and Russian scientists and even peasants had also discovered this technique and developed a theoretical understanding of it long before Lysenko. Since Lysenko referred only to a small part of this earlier work and only in his first publication on it, his advocacy of this technique as his own represented a kind of plagiarism.[48]

Nevertheless, Lysenko managed to convince the Soviet press and Soviet leaders that this was his idea and that it would greatly benefit agriculture. Because at this phase, so little was known about genetics, Lysenko and many other specialists and observers believed that this technique of artificial vernalization actually changed the "inheritance" of the plants and represented an example of the inheritance of acquired characters.[49] For Lysenko and his followers, artificial vernalization was thus a justification for their commitment to what they called "Neo-Lamarckism," which is usually viewed as a mistaken evolutionary theory that preceded Darwin and claims that organisms can pass on to their offspring characteristics they acquired during their lives.[50] Lysenko and some of his followers explicitly defended Lamarckism and Neo-Lamarckism at the infamous VASKhNiL conference in July–August 1948.[51]

In fact, Lysenkoists misunderstood what they observed. Recent genetics research shows that grain and other plants have specific genes that regulate their responses to changing temperatures and photoperiods, which these studies call VRN genes. Plants have multiple VRN genes and different versions or alleles of them, and their relationships can determine plants' reaction to vernalization.[52] The VRN genes also interact with other genes, making plants' ability to mature with or without a cold spell the result of complex polygenic interactions.[53] Thus the patterns that Lysenko and others saw in their artificial vernalization experiments were, in virtually all cases, the result of the plants' genetic potential rather than of any "change" of winter into spring wheat, both because of the complexity of plants' responses to vernalization and because significant genetic changes are rare events.[54] Lysenko's artificially vernalized plants may also have grown because of overlooked environmental conditions that had nothing to do with any change in the plants themselves.[55] Lysenko and his followers thus misinterpreted their experiments with winter and spring wheat by jumping to the conclusion that they had changed what they called the "inheritance" of the plant, when they actually observed variations among individual plants in their genetic make-up that affected their responsiveness to changing conditions.

Yet Vavilov and other scientists in the USSR saw artificial vernalization as potentially very useful (and initially supported Lysenko, despite his ignorance, for this reason). Vavilov and his colleague Nikolai Maksimov in 1932 argued that artificial vernalization could help breed plants with shorter growth periods. Vavilov also argued that the technique could allow the use of tropical and subtropical varieties in breeding by allowing crossing of varieties that normally had entirely different growth patterns. In 1933, Vavilov referred to grain varieties from Spain and North Africa that ripened in Saratov experimental plots to assert that the "simple technique of vernalization" allowed southern varieties to produce normal harvests in northern regions where they ordinarily could not ripen. Most notably, in 1934, Vavilov argued that artificial vernalization would facilitate the use of the enormous array of varieties in VIR's world collection in crossings to make new varieties.[56]

Later research and breeding practices have vindicated Vavilov's prediction. Artificial vernalization is still used as technique to transfer genetic characteristics between winter and spring crops. At CIMMYT, the Green Revolution center in Ciudad Obregón, Mexico, plant breeders routinely use "vernalization growth chambers" to allow "winter x spring crosses" which allow the transfer of important genetic characteristics between plants of different growth habits.[57] In this case, therefore, the technique that Lysenko advocated was actually legitimate and useful, even though he misunderstood it and caused pointless confusion and worse by imposing his interpretations of these processes on Soviet science.

Lysenko damaged but did not destroy the field of genetics in the USSR. Despite the tragic and unnecessary loss of Vavilov and several other leading scientists, the projects that Vavilov began resumed after the war. Some Lysenkoist ideas resulted from ignorance that would not be overcome for decades even in the west. Lysenko also subscribed to the same goals as the Soviet government and the scientists of the 1920s and before—to increase yields and production and protect the country from famine—even though he claimed that genetics and geneticists could not achieve these goals. It is unclear what Lysenko actually thought: Roll-

Hansen describes Lysenko as not only "sincere in his beliefs" but also refers to his "egotism." He cites a Soviet scientist who knew Lysenko and described him as a cynic who would "run down everything and everyone who obstructed his purpose."[58] These and other descriptions suggest that Lysenko may have been a sociopath willing to victimize others to advance himself.[59] This also implies that his crank ideas were, for him, a means to achieve power in science by deceiving Soviet officials. This suggests that scientists who did their work in ways that would not attract his notice or provoke his fear or hostility could continue to do legitimate work. At least one scientist managed to do this in a dramatic way.

Lukianenko and the Soviet Green Revolution

Pavel Pateleimonovich Luk'ianenko was the Soviet scientist who navigated the shoals of Lysenkoism and achieved a great deal using conventional genetics-based plant breeding during Lysenko's ascendancy. For the Soviet Green Revolution, he played a role comparable to Strampelli in Italy or Borlaug in the Mexican plant breeding program.

Luk'ianenko was born in Stanitsa Ivanovskaia, not far from present-day Krasnodar, to a Kuban Cossack family in 1901.[60] He not only farmed but also gained basic education before the war. He endured famine and deprivation during his childhood (including the famine of 1911), and then again during the world war and the civil war. His own and his family's experiences led him to side with the Bolsheviks. He served briefly in the Red Army after the Civil War and then enrolled in the Kuban agricultural institute in Krasnodar in 1922.

Luk'ianenko decided to become a plant breeder or *selektsioner* based on certain early life experiences. When he was an adolescent, a village elder described to him the effects of a plant disease the peasants called "zakhvat" that blackened wheat stems and allowed the plant to produce only depleted, "empty" grain. This infestation was stem rust, and this description of it was his first encounter with one of the plant diseases that would be a major focus of his plant breeding work.[61] At the agricultural institute, one of his teachers described the difficult but potentially very rewarding life of a plant breeder. This career required intelligence, stamina, and persistence, not to mention years of work to produce new varieties.[62] Luk'ianenko was, by this time, quite familiar with the chronic problems in Kuban farming, including rust and other plant diseases, lodging, droughts, and the skepticism and resistance of local peasants to scientific advice. The prospect of creating crop varieties to overcome these problems persuaded him to commit his life to this career in the 1920s.[63] During that period, he also married Polina Aleksandrovna, a fellow plant science student, who collaborated with him in this work.[64]

After graduating, the Luk'ianenkos worked at research institutes in Crimea and Chechnya, and in 1930, returned to Kuban.[65] During these years, they increasingly focused on developing varieties resistant to rust, lodging, and the extremes of cold and dry weather. The Kuban institute had a variety called "dvuruchka" that was both a spring and a winter wheat and was very resistant to cold, but not rust.[66] The Luk'ianenkos' breeding efforts expanded from local varieties to varieties from remote parts of the USSR such as Central Asia, and from abroad, including India, Germany, Canada, the USA, and Argentina. They obtained most if not all of these varieties from the collections gathered by Vavilov and his personnel at VIR.[67] Aleksandr Fedorchenko, Luk'ianenko's biographer, repeatedly states that Lukianenko knew and greatly respected Vavilov's work.

The Luk'ianenkos' breeding efforts were motivated by their own knowledge of chronic problems of Kuban wheat varieties, but collectivization and the Soviet famine made these problems even more urgent. With farm collectivization, the Soviet government commissioned plant specialists to develop varieties that could withstand machine harvesting. Such varieties should ripen simultaneously and be resistant to lodging and shattering (bursting of the ear and scattering of grain during harvesting), both chronic problems for grain growers in the Kuban and elsewhere in the USSR (as well as other countries).

The famine of 1931–1933 also resulted fundamentally from another set of environmental problems with which Luk'ianenko was familiar, but which neither Stalin nor most observers at the time and since have understood. Stalin did admit that crops in many areas of the USSR endured a severe drought in 1931, and Soviet authorities returned procured grain back to regions that suffered from that drought.[68] Yet Stalin in a speech in January 1933 stated that the 1932 harvest was not reduced by a crop failure because there was no major drought in 1932. Based on similar considerations, numerous publications have claimed that the famine was "man-made" or even genocide.[69]

In fact, however, the USSR did have serious crop failures in 1932 caused by environmental factors. Agronomists described and analyzed a series of infestation of crop diseases, insects, and rodents. The most significant was a vast and severe infestation of rust. Agronomists estimated that in 1932 rust caused losses of more than seven million tons of grain. In an extremely important study of the rust infestation in the Northern Caucasus in 1932, Luk'ianenko estimated that leaf rust destroyed at least 25 percent of the winter wheat harvest in Kuban alone, the top wheat-growing region in the USSR. The other infestations also caused substantial losses.[70] This infestation of leaf rust began with the 1931 crop. It was caused by a new race of rust to which almost no local varieties of wheat were resistant. Luk'ianenko, very apprehensive about the effects of this new rust, undertook to find or create wheat that would resist it. He crossed Soviet varieties, especially Ukrainka, with North American varieties including Marquis and Kitchener. Yet this rust strain caused the partial or complete failure of almost all of his 49 different hybrid crosses at the Krasnodar experimental station in 1931–1932.[71]

Luk'ianenko and the other agronomists at his institute worked intensively on breeding rust-resistant varieties in the wake of the 1931–1933 crises. By 1937, they produced 10,000 tons of hybrid seeds of wheat that were resistant to rust and cold, many of which could be planted both in fall and in spring. This seed stockpile may have been a factor in the large harvest of 1937. These plant breeders also continued to focus on short-stemmed varieties to prevent lodging, with larger consequences a few years later.[72]

World War II disrupted this work when the Nazi invaders reached the Northern Caucasus. Luk'ianenko and his wife and staff had to pack up their main materials very rapidly and flee the invaders. In the process, the Luk'ianenkos lost one of their children, a son, captured and killed by the Nazis in 1943.They also had to transport substantial amounts of grain, mostly hybrids they had produced over the previous decade, avoiding Nazi ground attacks and bombing raids, and then maintain their stocks while in exile in Central Asia. Polina Aleksandrovna was mostly responsible for this work, which resembles the story of the workers at VIR in Leningrad, who starved rather than eating the grain that Vavilov's expeditions had brought to the USSR.[73]

Upon the staff's return from wartime exile to the Krasnodar station in 1944, Luk'ianenko and his wife and associates resumed the projects of developing rust and cold-resistant crop varieties. They also conducted many experiments in planting winter wheats in spring and spring wheats in fall. These projects apparently reflected not only the influence of Lysenko, but also the limited understanding of genetics and the uncertainties about the effects of environment on inheritance that Roll-Hansen showed were prevalent at that time. According to Luk'ianenko's biographer, the breeder performed these experiments to find "universal" wheat varieties that could be planted both in fall and in spring, with the goal of having high-yielding spring varieties that could replant winter wheat fields destroyed by winterkill. They also continued cross breeding Soviet and foreign varieties, including the USA and Argentine varieties, resulting in an early ripening variety, Skorospel'ka, that could yield four tons per hectare, which was three to four times the normal yield of the time. By this time, Luk'ianenko had compiled a list of 26 characteristics of the ideal wheat variety, perhaps influenced by Vavilov's list of characteristics of an ideal variety in his book on wheat breeding.[74]

The most important result of the station's breeding efforts and the breakthrough that began the Soviet green revolution, was Bezostaia-1, a semi-dwarf variety that has repeatedly been recognized as one of the best winter wheat varieties ever produced.[75] Bezostaia means "awnless," and there was at least one natural awnless variety of wheat found in the 1920s in the Northern Caucasus. This may have been involved in the development of Bezostaia-1.[76] The breeding steps that led to this variety began in 1935 when Polina Aleksandrovna crossed Kanred-Fulcaster, an American variety, with Klein-33. Klein-33 was a hybrid wheat from Argentina that derived partly from Spanish rust-resistant varieties and partly from Ardito, one of the short-stemmed varieties produced by Strampelli in the 1910s and 1920s using Japanese dwarf wheat.

Polina Aleksandrovna had to quit this work in 1952 because of her health, but in 1953, the original crossing produced Bezostaia-4, which was a semi-dwarf variety 110 cm tall, resistant to lodging, rust, and cold. The Soviet government authorized use of this variety in 1955, and by 1957, it was planted on 350,000 hectares. Meanwhile, Luk'ianenko and his staff had crossed this with other types of wheat and came up with an even better variety, which they called at first Bezostaia-4/1, and later changed to Bezostaia-1.This variety had a shorter stem and higher yields than Bezostaia-4, was rust and cold resistant, and had very good plasticity, which mean that it was a good basic variety for crossbreeding. According to a Hungarian scientist, many later crossings would have been impossible without Bezostaia-1 as a basis.[77]

Luk'ianenko predicted early on that Bezostaia-1 could become the main wheat type for the Northern Caucasus. It grew a heavier head of grain than previous varieties, but because it was a semi-dwarf variety with a sturdy stem, even when mature, the plant stood straight up like a broom.

This made it highly resistant to even the heaviest rain and strongest wind, and very easy to harvest with low losses. Usually, mature wheat in Kuban stood bowed over and had a strong tendency to break and fall onto the ground, or lodge, which could cause large harvest losses. Farms growing Bezostaia-1 obtained harvests routinely double the size of harvests from earlier varieties, up to six metric tons per hectare.[78] This yield is in the order of magnitude of highest yields achieved in Mexico at the peak of the Green Revolution in the 1990s. Even if this six tons per hectare came only from the best-supplied farms, and the others had yields only half that level, the yield of three tons per hectare was still the average wheat yield in Mexico in the 1960s and 1970s, during the expansion of the Green Revolution there.[79] On the basis of this work, Luk'ianenko was promoted to full KPSS membership without a candidate stage, and also made an active member of the Academy of Sciences.

Bezostaia-1 was planted on large areas: at least 13 million hectares (32 million acres) by the late 1960s in the USSR and Eastern Europe, as well as in Iran, Turkey, and in other arid regions. By 1972, it was reportedly planted on 18 million hectares (45 million acres).[80] Western scientists consistently noted its high yields and plasticity, and recognized Lukianenko as one of the major wheat breeders of the world.[81] At an international meeting on plant breeding organized by Norman Borlaug in 1971, one questioner asked about "the Russian variety which yielded well at high latitudes in Turkey in the international winter wheat trials." Virgil Johnson and Norman Borlaug answered as follows:

V. A. Johnson: This winter wheat, Bezostaia, has been the highest yielding variety in the international winter wheat performance nurseries since the project was established in 1969. It is in a performance class by itself and it has wide adaptability. Morphologically, it is similar to the CIMMYT wheats.

N. E. Borlaug: Although this variety was developed in a local program, it has tremendous yield stability built into it.[82]

Clearly, Bezostaia-1 demonstrated that Soviet wheat breeding could reach the highest Western scientific standards. Perhaps the nearest comparison among Green Revolution varieties was the "Miracle Rice" IR8, bred by Henry Beachell and his coworkers at the International Rice Research Institute (IRRI) and widely distributed throughout Asia.[83] Connections between the Soviet and Mexican breeding programs grew. Scientists from Mexico had visited the Krasnodar institute.[84] In 1971, after Borlaug was awarded the Nobel Prize, *Literaturnaia Gazeta* asked Luk'ianenko to write about the Green Revolution, and Luk'ianenko praised Borlaug as an inexhaustible worker and organizer.[85] Lukianenko met Johnson at an international conference in Turkey in 1972. In February 1973, Borlaug wrote to Luk'ianenko, apologizing for missing the Turkey conference, and invited him to CIMMYT, but Luk'ianenko's heart condition prevented him from travelling. Luk'ianenko died in June from a heart attack while traveling around fields in Kuban to observe the growth of new wheat varieties.[86]

Luk'ianenko, Lysenkoism, and Soviet Genetics

Certainly, Lysenko and his associates harmed many legitimate Soviet scientists and misled Soviet biology students for several years. In this discussion, I do not intend to minimize these losses. Yet clearly, Luk'ianenko's work was not "thirty years behind" the leading Western workers in applied genetics.

Luk'ianenko and his associates at the Krasnodar institute bred semi-dwarf HYVs of wheat in the late 1940s and 1950s, at the peak of Lysenko's dominance, to create varieties that would be resistant to diseases, to extreme weather conditions, and to lodging. These were all the same objectives that Borlaug and his associates pursued in their work in Mexico at the same time or later, using some of the same basic varieties, including Strampelli's from Italy and Klein's from Argentina. Luk'ianenko's work was based on established concepts of genetics and inheritance. None of the Western specialists refer to him or his work in a negative way or as compromised by Lysenkoism. Luk'ianenko clearly saw his work as part of an international effort. He consistently used varieties from outside the USSR in his breeding, seeking new genetic material from them. His publication on the rust infestation of 1932 had a title page and section and table headings in English as well as Russian. Perhaps this was routine for agronomic publications at the time, but it also seems to reflect a desire to make this information available to Western researchers.

Where was Lysenko in all of this? Luk'ianenko's biographer repeatedly wrote that Luk'ianenko greatly respected Vavilov's work and did not mention Lysenko at all. In 1989, however, five years after the publication of that biography, the Soviet journal *Molodaia Gvardiia* published a long interview with I.A. Benediktov, Minister of Agriculture from Stalin to Khrushchev (1938–1958).[87] In this interview (about halfway through), Benediktov responded to a question about Lysenko as a charlatan in part by saying: "A dedicated student of Lysenko, who esteemed him to the end of his days, was Pavel Panteleimonovich Luk'ianenko, who was perhaps our most talented and productive *selektsioner*…" Benediktov named several of the wheat varieties that Luk'ianenko created, and also some bred by Lysenko, and added: "however one may criticize Lysenko, the cropland of our country to this day has dominant agricultural crops introduced by his students and people who sided with him." Soyfer, a harsh critic of Soviet agricultural policy, in a one-paragraph biography of Luk'ianenko in his book on Lysenkoism, wrote that Luk'ianenko "[R]ose to high administrative position in agricultural science by toadying to Lysenko, but then left administrative work to take up wheat selection in earnest."[88]

At this point I cannot document directly Luk'ianenko's attitude toward Lysenko, and sources for determining his attitude may not exist. Yet, according to Fedorchenko's biography, Luk'ianenko spent virtually his entire career in Krasnodar (he seems to have been absent only during the war), never studied with Lysenko, worked "in earnest" on wheat breeding throughout his career, and does not appear to have held any high administrative position. Luk'ianenko's crucial study of the 1932 rust infestation in the North Caucasus cites works by Vavilov, Artur Iachevskii, one of the founders of Russian and Soviet plant pathology, and L.F. Rusakov, another leading plant pathologist, but no publications or ideas from Lysenko.[89] The breeding of Bezostaia-1 began in 1935, before Lysenko had reached any significant position of control in Soviet science.

Perhaps something of Luki'anenko's attitude toward Lysenko can be inferred from the two articles he published in Lysenko's journals: one in *Iarovizatsiia* in 1941, after Lysenko had taken over VASKhNiL, the other in *Agrobiologiia* in 1948, at the peak of Lysenko's dominance.

The earlier article, "On the methodology of breeding winter wheat varieties resistant to leaf rust," examined how rust resistance in wheat varied in relation to environmental conditions. The article cited American studies and its research data employed American varieties, which Lysenko would not have favored. In the article's last paragraph, discussing the process of selecting a new variety for large-scale testing, Luk'ianenko wrote the only reference in the whole article to Lysenko: "At the same time we use the intravarietal crossing by means of castration and wind pollination proposed by Academician T.D. Lysenko."[90] This referred to one of Lysenko's irrational schemes to use open-air pollination instead of the controlled breeding that Luk'ianenko and other scientific breeders used. Luk'ianenko followed this with the last sentence of the article that referred to "maximum selectivity in fertilization" to present resistance to rust, completely the opposite of Lysenko's method that he mentioned in the previous sentence. It appears that Luk'ianenko cited Lysenko in this minimal, perfunctory, and dismissive way because he had to for some reason, perhaps as a gesture to Lysenko in return for publishing the article. Lysenko's approach was not an important part of Luk'ianenko's research, if he even used it at all; the article did not cite or list any publication or research by Lysenko. The article clearly was not the work of a "toady" or a "devoted student" of Lysenko.

The title of the second article, "Changing the nature of varieties of winter and spring wheat by means of changing the conditions of the process of vernalization," appeared to follow Lysenko's theories because it refers to a neo-Lamarckian approach of changing these plants' "nature" through inheritance of acquired characteristics from artificial vernalization.[91] In this article, Luk'ianenko discusses the results of experiments in planting winter wheat in spring and vice versa, which as noted above was one of the experiments that Lysenko misinterpreted. In passing Luk'ianenko refers to Lysenko's interpretation of "destabilized heredity." Yet this article also honestly and clearly reports how most of these experiments either failed or had limited success. More important, most of the article was oriented toward Luk'ianenko's main project of finding varieties resistant to rust and other threats, with multiple tables detailing the extent of rust resistance of every different spring-planted winter variety line. While this article seems to accept, at least implicitly, Lysenkoist claims such as "destabilized heredity" and the idea that planting winter wheat in spring "changed the nature of the variety," it also presented (as typical for Luk'ianenko) much straightforward evidence on disease resistance that had practical application for planting and breeding. It should be noted that agronomists in other countries have crossed winter and spring wheat and planted wheat varieties in the wrong season—in both cases, to find varieties resistant to cold spells in spring and unusually mild winters.[92]

Luk'ianenko may have cited Lysenko as "protective coloration," as Douglas Weiner described loyal statements conservationists made to distract the Soviet regime from their real values and motives Elina found examples of other plant breeders who evaded Lysenko's attention and thereby his control.[93] Perhaps when he had to deal with Lysenko or his associates, Luk'ianenko played the role of a dedicated follower of Lysenko, and thereby misled Benediktov. In his work, however, he remained in Krasnodar, some 1200 km from Moscow, and bred his wheat varieties using the standard genetics-based methods employed in the West. Luk'ianenko's 1948 article can also be seen as an example of the scientific manipulation and evasion of Lysenkoism described by Krementsov. Luk'ianenko's article in *Agrobiologiia* is, in part, "Lysenkoist" research, but it presents its evidence honestly, and contains potentially useful data from additional experiments that were Luk'ianenko's own focus and not strictly speaking "Lysenkoist" in orientation. Luk'ianenko's article could be seen as a "quasi-Lysenkoist" article, written with sufficient scholarly integrity to imply relatively clearly to an alert reader that a Lysenkoist approach was inadequate. We should note, finally, that Luk'ianenko's many other publications, with one exception of an article that referred to "Michurinism," never cited Lysenko or used his ideas.

Conclusions

The work of Luk'ianenko and certain other grain breeders led to an explosion of research on HYVs in Russia from the 1960s onward, as well as substantial genetics research and greatly improved education that began even before Lysenko's removal from power. These topics, however, lie outside the scope of this chapter.[94] This study of Luk'ianenko challenges the prevailing view that Lysenko held back Soviet genetics for a generation. While certainly, during the heyday of Lysenkoism, the Soviet regime victimized many excellent Soviet geneticists and wasted money and time on fraudulent Lysenkoists' "research," many other scientists conducted valid, substantial, and important work in this period—particularly in the area of plant breeding. Luk'ianenko was not the only agricultural scientist who did such research in these years, but his work had more national and international significance than that of any other Soviet agricultural scientist in this period.

This work differed greatly from the conventional view of scientific research in the time of Lysenko. Luk'ianenko's work began before Lysenko's rise and continued despite his dominance. Luk'ianenko's work relied substantially on plant varieties from outside the USSR, in most cases brought into the USSR through the work of Nikolai Vavilov. He also relied on conventional principles of genetics, including the guidelines for plant breeding published by Vavilov, as well as some breeding theories and techniques from outside the USSR. In particular, he and his coworkers independently sought and achieved the same goals as the Italians around Strampelli in the early twentieth century and Borlaug in the Green Revolution of 1950s–1960s.

Luk'ianenko's work during Lysenko's time and afterwards produced several extremely important wheat varieties that had the same characteristics as the Green Revolution varieties created by Borlaug. Luk'ianenko's Bezostaia-1, a semidwarf rust-resistant HYV, earned the highest praise from European and American breeders, including Borlaug, as one of the best of the HYVs. This finding thus goes beyond even Krementsov's points about scientists' evasion of Lysenko. Luk'ianenko and his colleagues, more than simply continuing previous genetics-based work in plant breeding, achieved breakthroughs that put their work at the forefront of world wheat breeding, both in their methods and their results. Because of these and other considerations, a post-Soviet Russian symposium on breeding of wheat and triticale commemorating Lukianenko was entitled "The Green Revolution of P.P. Luk'ianenko."[95]Thus, despite the repressive and irrational actions of Stalin and Lysenko, Luk'ianenko fulfilled the potential of Vavilov's writings and collection of international varieties, and as a result, Soviet agronomists and agriculture thus participated in the international Green Revolution.

Notes

1. See especially Mark Tauger, "Stalin, Soviet Agriculture, and Collectivization," in Trentmann and Just, eds., *Food and Conflict in Europe in the Age of the Two World Wars*, New York and Basingstoke: Palgrave MacMillan, 2006, 109–242, and Mark Tauger, "Modernization in Soviet Agricuture," in Kangaspuro and Smith, eds., *Modernization in Russia since 1900*, Helsinki: Finnish Literature Society, 84–103. Both of these can be read at the website: http://history.wvu.edu/faculty/current-faculty/mark-b-tauger/soviet-agriculture-and-famines.
2. Valerii Soyfer, *Lysenko and the Tragedy of Soviet Science* (Camden: Rutgers University Press, 1994), has a one-paragraph, somewhat inaccurate biography of Luk'ianenko in an appendix (p. 306) containing brief biographies of historical figures in the history of Lysenkoism. In this paragraph, Soyfer wrote: "He [Luk'ianenko] could well be placed in the category of people who laid the foundations of the

'green revolution,' which so greatly increased crop yields." Luk'ianenko did more than lay the foundations for the Green Revolution, as will be discussed below.

3. This discussion relies on John Perkins, *Geopolitics and the Green Revolution* (New York, 1997), Lennard Bickel, *Facing Starvation* (New York, 1974), and Susan Dworkin, *The Viking in the Wheat Field* (New York, 2009). My focus here is on wheat; the pattern for the Green Revolution in rice was similar, but involved the International Rice Research Institute (IRRI) and the work of the American rice breeder Henry Beachell.
4. Katerina Borojevic, Kzenija Borojevic, "The Transfer and History of 'Reduced Height Genes' (Rht) in Wheat from Japan to Europe," *Journal of Heredity* 2005 (96) 4:455–459.
5. The excellent popularized history by Susan Dworkin does discuss part of this earlier history.
6. Borojevic and Borojevic, "The Transfer and History of "Reduced Height Genes" (Rht) in Wheat from Japan to Europe," 455–459.
7. Dworkin, *Viking*, 38–39.
8. Gian Tommaso Scarascio Mugnozza, "The contribution of Italian wheat geneticists: From Nazareno Strampelli to Francesco D'Amato," in R. Tuberosa et al., eds., *Proceedings of the International Congress:" In the wake of the Double Helix: From the Green Revolution to the Gene Revolution," May 2003*, (Bologna, Italy: Avenue Media, 2005).
9. Roberto Lorenzetti, *Wheat Science: The Green Revolution of Nazareno Strampelli* (Rome: Journal of Genetics and Breeding, 2000), 120–122.
10. Lorenzetti, *Wheat Science*, 112 ff.
11. Mark Tauger, "Famine in Russian History," *The Supplement to the Modern Encyclopedia of Russian and Soviet History* [SMERSH] v. 10 (2011), 79–92.
12. Tauger, *Agriculture in World History* (London: Routledge, 2011), 103.
13. A recent excellent study is Ol'ga Elina, *Ot Tsarkikh sadov do Sovetskikh polei: Istoriia sel'skokhoziaistvennykh opytnykh ucherzhdenii XVIII-20-e gody XX veka*, 2 vols., (Moscow: Russian Academy of Sciences, 2008).
14. On Vavilov, see Peter Pringle, *The Murder of Nikolai Vavilov* (New York: Simon and Schster, 2008) and numerous Russian-language studies and sources, some to be cited below.

15. Valerii Soyfer, *Lysenko and the Tragedy of Soviet Science*, 44–48, and Vavilov, *The Origin, Variation, Immunity and Breeding of Cultivated Plants: Selected Writings of N.I. Vavilov*, translated by K. Starr Chester (Waltham: Chronica Botanica, 1949).
16. S.R. Mikulinskii, ed., *Nikolai Ivanovich Vavilov: ocherki, vospominaniia, materialy*, (Moscow: Nauka, 1987), 39. The publications include: S.M. Bukasov, *Vozdelyvaemye rasteniia Meksiki, Gvatemaly i Kolumbii.* (Leningrad: Institute Rastenievodstva, 1930), 553 pp.; N.I. Vavilov, D.D. Bukinich, *Zemledel'cheskii Afghanista. Agricultural Afghanistan. Sostavlen po materialam ekspeditsii Gos. in-ta opytnoi agronomiii Vses. in-ta prikladnoi botaniki v Afghanistan*, (Leningrad: Institute Rastenievodstva, 1929), 610 pp.; N.I. Vavilov, *Pshenitsy Abissinii i ikhpolozhenie v obshchei sisteme pshenits: K poznaniiu 28 khromozomnoi gruppy kul'turnykh pshenits*, 236 pp. (Leningrad: Institut Rastenievodstva, 1931).
17. E.I. Kolchinskii, A.A. Fedotova, *Nauchnyi Sankt-Peterburg: Biologiia v Sankt-Peterburge, 1703–2008: entsikopedicheskii slovar'*, (St. Petersburg, Russia: Nestor-Istoriia, 2011), 83, 114. The compilers are both historians at the Institute of History of Natural Sciences and Technology of the St. Petersburg branch of the Russian Academy of Sciences.
18. See Lucile Brockway, *Science and Colonial Expansion: The Role of the British Royal Botanic Gardens*, (New York: Academic Press, 1979), and Richard A. Walker, *The Conquest of Bread: 150 Years of Agribusiness in California*, (New York: New Press, 2004), p. 111.
19. Tauger, "Famine of 1921–1922," *Encyclopedia of Russian History*, edited by James Millar; E.M. Khenkin, *Ocherki istorii bor'by sovetskogo gosudarstvo c golodom* (1921–1922), (Krasnoiarsk, 1988).
20. See Tauger, "Grain crisis or famine?" in Raleigh, ed., *Provincial Landscapes: Local Dimensions of Soviet Power*, (Pittsburgh: University of Pittsburgh Press, 2001); Tauger, "Stalin, Soviet Agriculture and Collectivization," in Frank Trentmann, Flemming Just, eds., *Food and Conflict in Europe in the Age of the two World Wars*, (Basingstoke, Palgrave MacMillan, 2006), and Tauger, *Natural Disaster and Human Action in the Soviet Famine of 1931–1933*, (Carl Beck Papers, Pittsburgh: REES, 2001). These are available at my website: http://history.wvu.edu/faculty_staff/current_faculty/dr_mark_tauger
21. Vavilov, *Polevye Kul'tury Iugo-Vostoka* (Petrograd: Novaia Derevnia, 1922), 7.

22. S.K. Chaianov et al., *Opytnoedelo Narodnogo Komissariata Zemledeliia RSFSR: rezultaty, organizatsiia, programmy i plan na 1927-28-1931-1932 gg.*, (Moscow: Novaia Derevnia, 1929), 13 ff.
23. On VASKhNiL's 35 subordinate institutes that resembled CGIAR, and on its bureaucratic character, see Nils Roll-Hansen, *The Lysenko Effect* (Amherst, New York: Humanity Books, 2005), 77–78, 90.
24. Pringle, *Murder*, 167; Vavilov, *Polevye Kul'tury*, 9–10.
25. Pringle, *Murder*, describes Vavilov's frequent travels to Europe and the US for collaboration, often with other Soviet scientists. These scientists often published articles and books about their travels and collaborative work; one article by the Soviet drought specialist N.M. Tulaikov on the huge Tom Campbell farm in Montana persuaded Stalin that large-scale farming, and hence collectivization, was possible; Tauger, "Stalin, Soviet Agriculture, and Collectivization," 129.
26. Vavilov, *Origin, Variation, Immunity and Breeding*, 170–314.
27. Vavilov, *Origin, Variation, Immunity and Breeding*, 257–258.
28. On dekulakization, see N.A. Ivnitskii, *Kollektivizatsiia i raskulachivanie* (Moscow, 1994); on collectivization as "serfdom," see Sheila Fitzpatrick, *Stalin's Peasants* (New York: Oxford University Press, 1994); on collectivization as part of a larger modernization program, see Tauger, "Stalin, Soviet Agriculture, and Collectivization," and Tauger, "Modernization in Soviet Agriculture."
29. Joravsky, *Lysenko Affair*, 77.
30. My main sources include Soyfer, *Lysenko and the Tragedy of Soviet Science*; Krementsov, Stalinist Science (Princeton: Princeton University Press, 1997); Roll-Hansen, *Lysenko Effect*; Ethan Pollock, *Stalin and the Soviet Science Wars* (Princeton: Princeton University Press, 2006); Joravsky, *Lysenko Affair*, Zhores Medvedev, *The Rise and Fall of T. D. Lysenko*, (New York:1969); Pringle, *Murder*.
31. Noël Kingsbury, *Hybrid: The History and Science of Plant Breeding* (Chicago, University of Chicago Press, 2009), 209.
32. Soyfer, *Lysenko and the Tragedy of Soviet Science*, xviii, 5–7, passim.
33. Joravsky, *Lysenko Affair*, appendix.
34. Joravksy, *Lysenko Affair*, 112–130.
35. Professor Steve McCluskey, personal communication, based on graduate work with Loren Graham.
36. Krementsov, *Stalinist Science*, 239–253.

37. Roll-Hansen, *Lysenko Effect*, 22–24. Neo-Lamarckism refers to theories alleging the inheritance of acquired characters.
38. Pringle, *Murder*, 26–28.
39. See Ethan Pollock, "From Partiinost' to Nauchnost' and Not Quite Back Again," *Slavic Review* v. 68 no. 1, Spring 2009.
40. Joravsky, among others, narrates the main examples of Lysenko's projects.
41. See Soyfer, *Lysenko and the Tragedy*, Ch. 13–15, for a survey of Lysenko's decline.
42. See Soyfer, *Lysenko and the Tragedy*, 53–59, and Roll-Hansen, *Lysenko Effect*, 56–58, 94–95, 135–137, 160–163.
43. See Soyfer, *Lysenko and the Tragedy of Soviet Science*, 135, for Vavilov's last meeting with Stalin. According to this source, it seems clear that Stalin was exasperated with Vavilov's attempt to explain rationally the inevitable lag in plant breeding, and Soyfer documents Lysenko's slavish and mendacious promises to Stalin of fast work.
44. Pringle, *Murder*, 155–156; Soyfer, *Lysenko and the Tragedy*, 134–140.
45. Mark Popovsky, *The Vavilov Affair*, (Hamdon, Connecticut: Archon, 1984), 165–178; Pringle, *Murder of Nikolai Vavilov*, 266–279.
46. I.G. Loskutov, *Vavilov and his institute: A history of the world collection of plant genetic resources in Russia* (Rome: IPGRI, 1999), 124–132.
47. I.G. Loskutov, *Vavilov and his institute: a history of the world collection of plant genetic resources in Russia* (Rome: IPGRI, 1999), 117, 119, 124, 127, 130. The book mentioned is *Kul'turnaia flora SSSR. Mnogoletnye bobobye travy* (Moscow & Leningrad, 1950), 526 pages.
48. See Soyfer, *Lysenko and the Tragedy*, 12–20, Roll-Hansen, *Lysenko Effect*, 29–32, 55–57.
49. For example, Lysenko described the results of an experiment in planting winter wheat in spring in 1936 and 1937, in which the spring-planted varieties produced very few maturing plants in 1936 and even fewer in 1937, as having changed the plants' heredity. Lysenko, "Heredity and its Variability," in Lysenko, *Agrobiology* (Moscow, 1954), 427–428.
50. Roll-Hansen explains Lysenko's commitment to neo-Lamarckism and the viewpoint that environmental events could change plants' heredity, *Lysenko Effect*, 165–173. Apparently, the concept of Lamarckism, especially as opposed to "Darwinism," is a myth, since such views were very widespread at least in the nineteenth century,

and even Darwin accepted some of them. See Roll-Hansen, *Lysenko Effect*, 22–24, and Michael Ghiselin, Research Fellow at the California Academy of Sciences, "The Imaginary Lamarck," *The Textbook Letter*, September/October 1994, online at: http://www.textbookleague.org/54marck.htm.

51. Lysenko's speech is in *The situation in biological science; proceedings of the Lenin Academic of Agricultural Sciences of the USSR* (Moscow, 1948), and is available online: http://www.marxists.org/reference/archive/lysenko/works/1940s/report.htm.
52. Winter habit can result from an epistasis or interaction of vernalization or VRN genes at different loci in the genome. The VRN-1a gene has two alleles, for spring habit and winter habit, and the spring habit allele is dominant. For at least some wheat varieties to have winter habit, all the loci or sites for this gene on the chromosomes must have the recessive, winter-habit alleles. Some plants even have multiple versions of the recessive allele. A.E. Limin, D.B. Fowler, "Developmental Traits Affective Low Temperature Tolerance Response in Near-isogenic Lines for the Vernalization Locus VRN-1a in Wheat (triticumaestvum L. emThell), *Annals of Botany* 89, 2002, 579–585; A. T. Pugsley, "A genetic analysis of the spring-winter habit of growth in wheat," *Australian Journal of Agricultural Research*, 22 (1), 21–31, 1971.
53. Winter and spring habits can be polygenic traits in which several genes interact. Some studies show the existence of multiple VRN genes (VRN-2a, VRN 4, and so forth) that affect a plant's need for a period of cold, or vernalization, for flowering, while genes that determine a plant's response to photoperiod also affect habit. See among several works, Y.Y. Klaimi, C.O. Qualset, "Genetics of Heading Time in Wheat (Triticumaestivum L.). II. The Inheritance of Vernalization Response, *Genetics*, 76, January 1974, 119–133; D. K. Santra et al., "Genetic and Molecular Characterization of Vernalization Genes *Vrn-A1*, *Vrn-B1*, and *Vrn-D1* in Spring Wheat Germplasm from the Pacific Northwest Region of the USA., *Plant Breeding*, 128 (6), December 2009, 576–584.
54. According to Martin A.J. Perry et al., "Mutation discovery for crop improvement," *Journal of Experimental Botany*, 60 (10) 2009, 2817–2825: "Whilst mutations occur spontaneously in nature, the frequency of such mutations is too low to rely on alone for accelerated plant breeding." Many other scientific publications document the rarity of major genetic changes and mutations. Some of these varieties may have been facultative wheat, which has less of a vernalization response and can be planted in both fall and spring; Reinhard Neugschwandter et al.,

"Development, growth, and nitrogen use of autumn- and spring-sown facultative wheat," Acta Agriculturae Scandinavica, Section B Soil and Plant Science, 65 (1), 2015, 6–13, http://www.tandfonline.com/doi/abs/10.1080/09064710.2014.958522?journalCode=sagb20 . Facultative grain varieties were discovered in the late 19th-early 20th centuries by Russian scientists, but I have found no mention of them in discussions of Lysenkoism.
55. Artificial vernalization produces sprouted seed that will grow into a plant if, for example, it is planted before mold starts to grow on the seedlings. One website describes a method to sprout wheat berries to produce wheatgrass that resembles Lysenko's artificial vernalization, and warns about the spread of mold if the sprouts are not removed from the sprouting process in time: http://wheatgrassgrower.blogspot.com/2009/11/planting-your-wheat-berries-to-sprout.html. Also, winter wheat planted in spring, for example in March or April, may become cold enough in the soil, or because of a cold spell, to mature without any change in "inheritance."
56. Roll-Hansen, *Lysenko Effect*, 159–162.
57. See, for example, the image and mention of a vernalization growth chamber (actually a type of refrigerator) at CIMMYT, used to grow winter wheat to develop resistance to the rust epiphytotic Ug99: http://www.flickr.com/photos/cimmyt/4809668488/. See also A.I. Morgounov et al., *Wheat Breeding: Objectives, Methodology, and Progress*. Proceedings of the Ukraine/CIMMYT Workshop, June 1995, Wheat Program Special Report, WPSR no. 37, (Sonora Mexico: CIMMYT, 1995), p. 11ff "Importance of winter x spring crosses."
58. Roll-Hansen, *Lysenko Effect*, 58, 73, 106.
59. See the discussion of the characteristics and widespread incidence of sociopathy in society in Martha Stout (professor of psychology, Harvard University), *The Sociopath Next Door* (New York: Random House, 2005). On Lysenko's psychology, see Soyfer, *Lysenko and the Tragedy of Soviet Science*, passim.
60. The following biographical discussion is based on Aleksandr Fedorchenko, *Luk'ianenko*, (M: Molodaia Gvardiia (Zhizn' zamechatel'nykh liudei), 1984); Vitalii Vardadym, "Pavel Panteleimonovich Luk'ianenko, 1901–1973," *Radeteli zemli kubanskoi*, http://kuban-xxi.h1.ru/favourite/508.shtml; I.N. Elagin, "Vydaiushchii uchenyi-selektsioner. K 80-le iiu so dnia rozhdeniia P.P. Luk'ianenko," *Zhurnal RAN*, 1981 no. 11, 104–111.
61. Fedorchenko, *Luk'ianenko*, 49–50.
62. Fedorchenko, *Luk'ianenko*, 104–105.

63. Fedorchenko, *Luk'ianenko*, 105ff describes several agricultural threats with which Luk'ianenko and his fellow farmers struggled in the early 1920s, including lodging from rain, plant diseases and insects, drought, heavy rains, and cold. Late imperial Russian publications even mentioned soil exhaustion in Kuban (108).
64. Fedorchenko, *Luk'ianenko*, 111.
65. Fedorchenko, *Luk'ianenko*, 113–115.
66. Fedorchenko, *Luk'ianenko*, 114–115. This may have been a facultative wheat variety.
67. A.A. Romanenko, "Bezostaia 1 - triumf nauki i iskusstva." *Bezostaia 1 - 50 let triumfa*. Sbornik materialov mezhdynarodnoi konferentsii, posviashchennoi 50 letiiu sozdaniia sorta ozimoimi agkoi pshenitsy Besostoi 1. (Krasnodar:2005), p. 8.
68. Iurii Moshkov, *Zernovaia problema v godys ploshnoi kollektivizatsii* (Moscow, 1966), 191.
69. I.V. Stalin, *Sochinenniia*, v. 13 (Moscow, 1954), 216–233. On the genocide interpretation, see for example, Robert Conquest, *Harvest of Sorrow* (New York, Oxford University Press, 1986), and for a critique showing that the 1932 harvest was in fact a small harvest, Tauger, "The 1932 Harvest and the Famine of 1933," *Slavic Review*, v. 50 no. 1, Spring 1991, 70–89.
70. Tauger, *Natural Disaster and Human Action*, 17; Lukianenko, *O stepeni ugneteniia gibridov ozimoi pshenbitsy buroi rzhavchinoi v 1932 g.* (Rostov na Dony, 1934), 14.
71. Elagin, "Vydaiushchii uchenyi-selektsioner, " 106.
72. Fedorchenko, *Luk'ianenko*, 122.
73. Fedorchenko, *Luk'ianenko*, 125–145.
74. Fedorchenko, *Luk'ianenko*, 149–152.
75. The following section is based on Fedorchenko, *Lukianenko*; Elagin, "Vydaiushchii uchenyi-selektsioner," Vardadym, "Pavel Panteleimonovich Lukianenko," and other sources.
76. P. Grebennikov, *Tverdaia bez'ostaia pshenitsa v estestvennykh usloviiakh*, RnD, 1925.
77. Fedorchenko, *Luk'ianenko*, 164.
78. Fedorchenko, *Luk'ianenko*, 165–172.
79. David B. Lobell et al., "Analysis of wheat yield and climatic trends in Mexico," *Field Crops Research* 94 (2005) 250–256, citing FAO data on Mexican wheat production.
80. Dworkin, *Viking*, 90.

81. Elagin, "Vydaiushchii uchenyi-selektsioner," 108.
82. Borlaug, N.E."Breeding wheat for high yield, wide adaptation, and disease resistance." In *Rice Breeding*, IRRI, Philippines, 1972, p. 590.
83. On Beachell's breeding of IR8, see http://www.livinghistoryfarm.org/farminginthe50s/crops_17.html and the pages that follow this one.
84. Fedorchenko, *Luk'ianenko*, 202.
85. Fedorchenko, *Luk'ianenko*, 202–204.
86. Fedorchenko, *Luk'ianenko*, 217–219, 235, 260–261. According to an interview with an American plant breeder, Luk'ianenko died after seeing a major new infestation, Dworkin, *Viking*, 50–51. This story is problematic because it describes Luk'ianenko as "at his home near the "Odessa Institute of Plant Breeding" but Luk'ianenko's home was in Kuban and he did his research at the Krasnodar institute, several hundred miles from Odessa. Perhaps this source confused Luk'ianenko with another person.
87. I.A. Benediktov, *O Stalin i Khrushchev*, *Molodaia Gvardiia,* 1989 no.4, 12–65.Obtained from the Internet: http://rksmb.org/get.php? 143.
88. Soyfer, *Lysenko and the Tragedy of Soviet Science*, 306.
89. Luk'ianenko, *O stepeni ugneteniia*, pp. 45–46.
90. P.P. Luk'ianenko, "O metodike selektsii sortov ozimoi pshenitsy, ustoichivykh k buroi rzha vchiny," *Iarovizatsiia*, 1941 no.3, pp. 38–47, quote from p. 46.
91. Luk'ianenko, "Izmenenie prirody sortov ozimoi i iarovoi pshenitsyputem izmeneniia uslovii prokhozhdeniia stadia iarovizatsii," *Agrobiologiia*, 1948 no. 2, 1948, 40–50.
92. See, for example, International Winter Wheat Improvement program (Turkey-CIMMYT-ICARDA)http://www.iwwip.org/files/iwwip-website-info.pdf, apparently from 2008 or later.
93. Douglas Weiner, *A Little Corner of Freedom* (Berkeley: University of California Press, 1999), 41 and passim; Ol'ga Elina, "Mezhdu nauchnoi teoriei i sel'skokhoziaistvennoi praktikoi. Selektsionery i Lysenko (1948-1955 gg.), Za "Zheleznym zanavesom": mify i realii sovetskoi nauki, ed. E. I. Kolchinskii et al., Sankt Peterburg: Dmitrii Bulanin, 2002, 376–392.
94. I am preparing a study of famines in Russia and the USSR that will address many of these issues.
95. *Pshenitsa i tritikale*: *materialy naucho-prakticheskoi konferentsii 'Zelenaia revoliutsiia P. P. Luk'ianenko'*, (Krasnodar, 2001) 799 p.

Lysenko's "Michurinism" and Art at the Moscow Darwin Museum 1935–1964

Pat Simpson

Introduction

The triumph of Lysenko's "Michurinist biology" in 1948 did not just affect Soviet biological sciences per se. It also profoundly affected the ways in which biological science and Darwinian evolutionary theory were presented for mass consumption in Soviet natural history museums. This chapter offers a case study from an art historian's perspective of one rather special example of such museums—the Darwin Museum in Moscow—in relation to the growth and decline of Trofim Lysenko's power between 1935 and 1964. The chapter focuses on the discursively contextualised use of artwork (including artistically taxidermised specimens) by the museum, as a means to defend and protect its position and that of its directorate in politically difficult times, and also to gain access to state resources.

The Darwin Museum was founded in 1907 by a young ornithologist and amateur taxidermist from a Russian-German family, Professor Aleksandr Fedorovich Kots. When Kots was appointed to teach evolutionary theory at the Women's Higher Courses Institute of Moscow University, he brought with him his extensive and constantly increasing personal collection of

P. Simpson (✉)
University of Hertfordshire, Hatfield, Hertfordshire, England

W. deJong-Lambert, N. Krementsov (eds.),
The Lysenko Controversy as a Global Phenomenon, Volume 1,
DOI 10.1007/978-3-319-39176-2_5

books, illustrations, and rare stuffed zoological specimens to use as visual teaching aids.[1] The collection was formally adopted by Moscow University in 1914, and by 1926, it was spread over three floors of one wing of the former Women's Higher Courses building.[2]

In designing the museum, Kots and his wife, the zoo-psychologist, Nadezhda Ladygina-Kots,[3] had a vision (which lives on at the museum today) of a natural history museum in which art works would play a crucial role in stimulating visitors to study natural science.[4] This vision was perhaps partly fed by their acquaintance with the imaginative, large-scale reconstructive mural of *Stone Age Feast* (1883–1886) by the Russian artist Viktor Vasnetsov in the Imperial Russian Historical Museum, Moscow. It was certainly further nourished by their honeymoon tour of European natural history museums in 1913, where they would have encountered such prominent artworks as Emmanuel Fremiet's dramatic sculpture of a *Gorilla Carrying off a Woman* (1887) at the Jardin des Plantes in Paris, and aspects of Aimée Rutot and Louis Mascrae's more scientific series of reconstructive sculptural busts representing early hominids (1909–1914) at the Institut Royale des Sciences Naturelles Belgique in Brussels.[5] These sources exemplified some of the types of artworks that Kots was to commission for the Darwin Museum between 1914 and 1921, mainly from the artist Vasilii Vatagin, who had been a fellow zoology student at Moscow University, and rapidly became one of the leading zoological illustrators of the Soviet state.[6]

After the Bolshevik Revolution of 1917, the museum was nationalised as an adjunct to Moscow State University,[7] overseen by the Commissariat for Enlightenment (Narkompros) headed by Anatolii Lunacharskii. By 1924, the Darwin Museum had shifted from a purely academic institution concerned with zoological research and the education of Moscow University students to a broader remit that included the education of the general public—mainly workers, soldiers, school teachers, and schoolchildren—and the popularisation of Darwinism. Surviving the threats and rigours of the Cultural Revolution 1928–1932, the backlash of the Stalinist purges of the mid-to-late 1930s, and the triumph of Lysenkoism in the late 1940s, Kots and his wife, remarkably, remained as directors of the museum from 1917 until their respective deaths in 1964 and 1963.

The new socio-political context for the museum after 1917 brought with it both opportunities and potential pitfalls. Funding from the new Soviet state, for instance, enabled Vasilii Vatagin, and the respected taxidermist, Filip Fedulov,[8] both of whom had been engaged with the museum project from 1907, to be paid for their post-Revolutionary museum work. Prior to 1917, they had provided their services to Kots free of charge, indicating that the pre-Revolutionary funding for the nascent museum had been very poor.[9]

The emphasis of the Soviet regime on educating the working classes, particularly in relation to evolutionary science, also opened up new possibilities of access to greater resources from the new funder that might enable the directors to gain maximum fulfilment of their ambitions for the museum displays. Their aspirations included the construction of a new and larger building to accommodate the huge and ever-expanding collection. According to Paul Kammerer, who visited the museum in 1926, the displays were by this time already cramped and crammed, implicitly needing more space.[10] In the same year, there was actually a government decree issued for the construction of a new building for the museum, but this promise was seemingly overtaken by the priorities of the first five-year plan and not fulfilled.[11] Although the museum was again promised a new building in the 1950s, due to other contextual factors this did not materialise in the Soviet era. However, this goal was indubitably of focal importance to the museum throughout the period 1917–1964, and hence a major influence on policy decisions.

The pitfalls of coping with the new socio-political context arguably lay in the specific shifts regarding the definition of Darwinism and its relationship with constructs of Marxism, developed and debated within the Soviet Union from the 1920s to the 1960s. During this period, Kots always seemed to ensure that the museum displays were politically correct in relation to the changing demands of the Soviet regime and the corresponding shifts in the language through which bioscience was publicly disseminated and debated. Indeed, it was in his interests to do so. From 1926 onward, the museum was faced with a constant barrage of inspections by the authorities, which could have led either to the museum's closure or to the removal of the directorate.[12]

Since art was so fundamental to the museum displays, it was also through art, often in terms of sculpture busts, portraits, and narrative paintings of scientific "heroes", that the museum both declared its allegiances, and hedged its bets in relation to contemporary scientific debates. Equally important here, as will be seen, was both the physical presentation of some displays and their discursive framing by Kots' lectures and speeches within specific cultural, political, and economic contexts.

Nothing better exemplified Kots' keen nose for political correctness regarding the museum's strategic use of artworks, than his lightning response in August 1948 to the decision of the All-Union Lenin Academy of Agricultural Sciences (VASKhNIL) to abolish "reactionary" genetic science. His fast footwork resulted in a plethora of paintings, some plaster sculpture busts, and new or reframed displays that were produced and exhibited within a very short time. These were arguably instrumental in giving a successful, "Michurinist" gloss to the museum's representation of Darwinism in the period dominated by Lysenko, thus securing the continued existence of the Darwin Museum under its original directorate.

In relation to this, the chapter also suggests that the Darwin Museum art commissions in 1948, as well as others made between the 1930s and late 1950s, viewed in conjunction with archival materials held at the Darwin Museum and in the UK[13] might be seen to some extent as an intriguing and unusually visual gauge for charting the gradual and by no means uncontested, rise and fall of Lysenko. The argument also suggests that Kots was enabled to protect the museum by tapping into the current and mutable versions of what the Russian historian of science, Nikolai Krementsov, has characterised as a public "cultural resource"—"Marxist Darwinism".[14]

Krementsov has convincingly argued that the 1920s Bolshevik campaigns to popularise Marxism and to present Darwinism as a properly "materialist" scientific discourse integral to Soviet education became combined with the concurrent debates initiated by Soviet geneticists opposed to the contemporary trajectory of Russian scientific engagement with neo-Lamarckian ideas. He has also argued that by the early 1930s, these discursive trajectories had converged to transform Darwinism into a "public cultural resource".[15] This public resource became increasingly distanced from the scientific notion of Darwinism as "a specialised, often esoteric knowledge of the laws and principles of evolution, heredity and variability".[16] According to Krementsov, it developed its own simplified polemical "lexicon",[17] was highly context-dependent, and as a result, was an infinitely moveable feast that could be used by interested groupings to assert their own positions and attack those of other groups. Alternatively, it could be used to demonstrate the allegiance of individuals and institutions to the version of "Darwinism" currently being upheld by the Communist party and state. Arguably, the Darwin Museum's engagement with Marxist Darwinism can be seen to have had the latter function.

At a specific level, this study can be seen to provide a new and vivid example of how the priority to engage with Marxist Darwinism operated in the period under review among those institutions and individuals who were not key players in the debates over genetics, but nevertheless, contributed, however, unwillingly, to the entrenchment of Lysenkoism. At a broader level, the study implicitly illustrates some of the attendant dangers of transforming the complex discourses of science into simplified and demagogic "cultural resources", in particular, the suppression of public access to the complexity and relativism of real scientific discourse.

The Darwin Museum and the Shifting Sands of "Marxist Darwinism"

As Krementsov has suggested, Russian translations of Engels' unfinished essay "The Role of Labour in the Origin of Man from Apes" (1922), and the book *Dialectics of Nature* (1925), played a significant role in the development of Marxist Darwinism from the early 1920s.[18] The first of these texts identified engagement in labour as the decisive factor in human evolution from apes.[19] *Dialectics of Nature* further emphasised Darwin's Malthusian "error" regarding the intra-species "struggle for existence" as being necessarily applicable to humans. Engels' critique of Darwin, building on Karl Marx's reservations about the Malthusian element in *Origin of Species*, had asserted that cooperation, not competitive "struggle", was the true path of human evolution, particularly after the envisaged social revolution.[20] These ideas were to be fundamental to the successful rhetoric of Lysenko's 1948 speech, endorsed by Stalin's editorial hand on the final draft.[21]

The gradual alignment of the Darwin Museum with Marxist Darwinism, and hence, Engels' ideas on evolution, can be seen to some extent in the 1920s. Perhaps fortuitously, some sculptures and paintings made by Vatagin during the very early 1920s implicitly depicted cooperation and labour as definitive elements in the successful evolution of primitive humans. This may have had more to do with the museum's affinity for a contemporary tendency within Russian and early Soviet Darwinian discourse, exemplified by Prince Piotr Kropotkin's construct of "mutual aid", to prefer notions of cooperation to that of "struggle for existence".[22] Nevertheless, the images could easily be discursively recontextualised in the later 1920s–1930s as aligned with Engels' ideas.

More specifically, during the Cultural Revolution, a treatise on the behaviour of macaques published in 1928 by Ladygina-Kots, emphasised contemporary monkeys' aversion to cooperative labour as indicating their "degeneration" from the primate progenitors of humankind.[23] Albeit without an explicit source reference to Engels, this work implicitly followed the contemporary trend in the construction of Marxist Darwinism, exemplified by the *Great Soviet Encyclopaedia*'s entries for "Struggle for Existence" (1927)[24] and "Darwin and Darwinism" (1930).[25] The impetus behind Ladygina-Kots' implicit allusion to Engels may have been the hope that the new building promised by the Decree of 1926, might yet be fulfilled.

The Darwin Museum's strategic adoption of aspects of the current, developing political lexicon of Marxist Darwinism was explicitly indicated by the emphasis, in a course outline by Kots in the 1930s, on the importance of Engels' views about evolution and Darwin.[26] The stress on Engels in Kots' text, and the raft of party publications cited in the "recommended reading for teachers", pointed to the strengthening hold of the party over Soviet bioscience in the mid-1930s. This situation is particularly highlighted by the polemical 1935 entry on "Evolutionary Theory" in the *Great Soviet Encyclopaedia* by V.M. Kaganov.

Kaganov was one of Lysenko's main supporters, and in this article, he concluded by asserting that engagement with the dialectical materialist approach as prescribed by Engels, was the "cure" for the contemporary "crisis" in biology and in evolutionary theory.[27] It is, therefore, significant that in 1936, Kots deemed it necessary to commission two busts of Engels from Vatagin, which may have been a response to the nature of the polemical debates sparked off by the debut of Prezent and Lysenko in the higher echelons of the Soviet scientific power structure between 1935 and 1936. At this point, the debate was increasingly characterised on both sides by reference to socialist political theorists perceived as authoritative in relation to Soviet biology.[28]

In this context, presumably, the sculptural busts of Engels were designed to function both as visual reinforcements of Kots' teachings, and also as tangible, visual signifiers of the Darwin Museum's "correct" contemporary presentation of Marxist Darwinism in the mid-late 1930s. As such, these sculptures were open at this time to interpretations as indicators of alignment, not only with the current interests of the Communist Party, but also, ambiguously, with those of both of the main camps in the contemporary genetics debate—the geneticists and the Lysenkoists.

Kots' notes on the structure of his "10 Hour Course on Human Evolution" for school teachers also show that by the mid-1930s, he was advocating criticism of racism, eugenics, and "social Darwinism" with particular reference to the "bourgeois American", Henry Fairfield Osborn.[29] This was very much in line with the tenor of Kaganov's *Great Soviet*

Encyclopaedia entry on "Evolutionary Theory" in 1935[30] in which both Osborn, a prominent eugenicist and director of the American Museum of Natural History (AMNH), and William King Gregory, a zoologist and secretary of the American Eugenics Society who also worked for that museum, had come under fire.[31]

Curiously, however, the sentiments expressed in the course outline implicitly conflicted with Kots' ongoing correspondence with Gregory regarding the delivery to the Darwin Museum of a bust of Osborn by the American sculptor Chester Beach, which eventually arrived between 1935 and 1936.[32] This bust represented the final element in a chain of correspondence that had started in the early 1920s with Osborn's selective fulfilment of Kots' requests for Western publications on bioscience that were unobtainable in the USSR.[33] In return for these, Kots had sent Osborn a bust of Francis Galton—the founder of eugenics—which was eventually displayed at the entrance to the 3rd International Eugenics Congress at the AMNH in 1932.[34] The course outline also conflicted with the ongoing correspondence between the Kotses and the American eugenicist and ape-researcher, Robert Yerkes.[35]

The presence of such contradictions, viewed in the light of Krementsov's model of Marxist Darwinism, suggests that by the mid-1930s, the Darwin Museum was utilising some of the polemical lexicon effectively set up in the 1920s and expanded in the 1930s, to signify the museum's political correctness, self-defensively, for public and state consumption. This strategy was arguably motivated in part to protect its own private trajectories of research interest in visually charting and staying abreast of the developments in international scientific discourse broadly relating to Darwinism.

In relation to Michurin, rather than "Michurinism," it would seem from Kots' personal library that he was well aware of Michurin's works through publications dated from 1936 to 1941, clearly acquired after Michurin's death on June 7, 1935.[36] While the museum itself had no image of Michurin before 1948, Vatagin, the museum's main artist, had made a 4 m-high concrete statue of him in 1936–1937 for the main gateway of the Moscow Zoological Gardens, with which the museum had research connections.[37] So there was at least a useful, if somewhat tenuous historical connection of the museum with acknowledgement of Michurin before 1948.

There was an equally tenuous early link between the museum and the idea of the primacy of habit, environment, and the inheritability of acquired characteristics in evolutionary development, which lay at the heart of

Lysenko's Michurinism.[38] This idea was rooted in the neo-Lamarckism rife in Russian and early Soviet scientific discourse. It was evident in the 1927 *Great Soviet Encyclopedia*'s extensive entry on "Struggle for Existence",[39] and had led, for instance, to the Communist Academy's invitation to Paul Kammerer to set up a research laboratory in Moscow in 1926, even while his research results were being analysed and doubted in the West.[40] This scheme eventually came to nothing due to Kammerer's suicide in the same year, but it is evident that he had a range of Soviet supporters, notably the Commissar for Enlightenment, Anatolii Lunacharsky.[41] It is possible that these supporters also may have included Aleksandr Kots at the Darwin Museum to which Kammerer made a visit in 1926,[42] and about which he subsequently published an enthusiastic review article before his death.[43] Kots himself had been dubious in the early 1920s about the connection between Mendel's ideas and those of Darwin forged by Russian geneticists such as Aleksandr Serebrovskii, and maintained a respect for Lamarck as an important precursor of Darwin.[44] Indeed, within the museum display, the only two monumental, full-portrait sculptures were (and still are) of *Blind Lamarck and his Daughters*, c.1921, and *Seated Darwin*, c.1927.

The sculpture of Lamarck represents the French scientist as an object of pity for his lamenting daughters—and by implication, also for the viewers. In the 1920s, particularly given Vatagin's training by the Russian Symbolist artists Konstantin Iuon and Ivan Dudin,[45] this image's symbolism may have been available for Soviet interpretation on two levels. On the one hand, it could be viewed entirely sympathetically as depicting Lamarck as an impoverished and ailing victim of the contemporary French monarchist scientific hierarchy in 1829, due to its domination by his enemy, the politically powerful natural historian Georges Cuvier. On the other hand, it could be seen a tribute to Lamarck's contribution to evolutionary theory, but implicit critique of his "blindness" to the conclusions that Darwin was later to draw. The latter mode of interpretation would have been more likely in the USSR from the 1930s onward, as "Lamarckism" increasingly became a term of serious abuse in the ensuing genetics debates.[46] This was certainly the view of Vatagin's "semi-symbolic" sculpture that Kots was to outline in his draft biography of Vatagin, written in the Lysenskoist context of 1952.[47]

What enabled Kots to reframe the artworks discursively in the Darwin Museum between the 1920s and 1948 in relation to the development of public discourse on Marxist Darwinism, was the deliberately "laconic" nature of the displays.[48] The low level of explanatory textual material provided therein mainly ensured that the museum's position in the contemporary debates on Darwinism could not be clearly identified by the viewer purely by looking at the display. Interpretation of the museum displays by the beholders was thus, largely constrained and directed by the nature of the discursive expositions offered to them by the museum tour leader, since there was no opportunity for visitors to browse on their own.

"Michurinizing" the Darwin Museum

In relation to the above discussion, it may be seen that the Darwin Museum held some materials that represented the likenesses or ideas of some scientists branded from the 1930s as "reactionary". Yet by 1948, the ways in which the displays were presented to Soviet visitors would have recognisably included aspects of the main elements of Lysenko's theorisation of Michurinist biology. This is not to say, however, that the Darwin Museum was fully prepared for the implications of Lysenkos' victory in July–August 1948.

On August 12, 1948, five days after the VASKhNIL session closed, the impact of Lysenko's speech on "The Situation in Biological Science" was formally registered at the Darwin Museum in papers given to the museum staff and to the Museum's Scholarly Council by Aleksandr Kots.[49] Kots was officially responding not to the VASKhNIL session per se, but rather, to a concomitant directive from the Committee for the Affairs of Cultural Enlightenment Institutions of the Council of Ministers USSR, which had already been discussed and inevitably endorsed by the museum directorate before the presentations were given.[50] Kots' carefully crafted speeches began by outlining the two immediately pressing issues for the museum, as handed down from above. These were first, to review the current displays and weed out any elements that could be interpreted as supporting "Weismannism (Mendelism-Morganism)"—that is to say, anything that might refer to what we now would label as "genetics"; second, to "discuss measures for the reconstruction" of the Museum displays in the light of the VASKhNIL session's outcomes.[51]

Kots' address to the Scholarly Council began by asserting that the Museum had no "reactionary materials".[52] While this was an obvious, self-defensive statement to make under the circumstances, it was not true. The private libraries and archives of Kots and Ladygina-Kots contained any number of works by and correspondence with the sorts of Soviet and foreign scientists specifically condemned by Lysenko in 1948.[53] Furthermore, the museum display of sculptural busts representing heroes of Darwinian science made by Vatagin in the 1920s–1930s included images of genetic experimenters and theorists such as Gregor Mendel; August Weismann; Wilhelm Johannsen; Hugo de Vries; and William Bateson—the inventor of the term "genetics".[54] All of these people had been identified by Lysenko in 1948 as "reactionary" interpreters of Darwin.[55]

This category also implicitly included other Western scientists, prominent for their championship of racism and eugenics, such as Galton, Yerkes, and Osborn, who, as already mentioned, were also represented in the Darwin Museum's collection of portrait busts.[56] Additionally, there was a bust of another of Lysenko's targets for abuse, the Russian eugenicist and geneticist, Nikolai Kol'tsov,[57] a work that had been donated to the museum by the famous Soviet sculptor Vera Mukhina in the late 1920s–early 1930s.[58]

But, as Kots had quite properly observed, images in museums are vehicles of ideology.[59] After August 1948, it was not that the names of these scientists could not be mentioned and images at least of some of them displayed,[60] as long as they were suitably framed for public exposure as representing a part of the historical development of Darwinism now rejected by the Soviet Union. One example of this thinking is in the draft biography of Vatagin written in 1952. Here, Kots defended the presence of the Weismann and Galton busts in the museum on the grounds that these scientists were legitimately part of the history of Darwinism, but had fallen into error (implicitly in relation to Lysenko's "Soviet Creative Darwinism") because, "like Darwin, they knew nothing of scientific socialism".[61] Kots thus also used an aspect of the Marxist Darwinist "lexicon" to justify the retention of some images which, for him, represented significant elements of the history of Darwinism.

Kots' August 1948 speeches focused on three "positive" actions to be taken to demonstrate the museum's full compliance with the required focus on Creative Darwinism set out by Lysenko in his exposition of Michurinism as the true scientific pathway for Soviet biology.[62] As Irina Kalacheva and Anna Kliukina have authoritatively observed, these actions were to produce a "radical" transformation of the displays and of their exposition within five months of the VASKhNIL session—that is to say, by December 1948.[63] There were also further tweakings of the displays between 1948 and 1955 to consolidate the museum's Michurinist credibility.

One aspect of the August 1948 plan was to commission monumental sculptural busts of Michurin, Lysenko (Fig. 2), and the Russian Darwinist most favoured by Lysenko, Kliment Timiriazev (Fig. 1).[64] These sculptures,

Fig. 1 V.A. Vatagin, *Bust of Kliment Timiriazev*, 1948, plaster, c.50 × 50 × 75cm, in A.F. Kots, "Vasilii Alekseevich Vatagin i ego raboty v Darvinovskom muzee 1902–1952", chapter iv, no date, AGDM, f.10141, o.623, ed.khr.215, photo album, p.15. By kind permission of the State Darwin Museum Moscow. Photo © author, 2010

and particularly the busts of Lysenko and Michurin, were probably produced in short order before the end of the year, possibly even before the end of the month, since the sculptor, Vatagin, was known for his ability to make such busts in a matter of days, working from photographs.[65]

The fact that the Darwin Museum did not already have a bust of Timiriazev speaks volumes for the potentially cutaneous and strategic nature of its engagement with the developing cultural resource of Marxist Darwinism which, by the late 1930s–1940s, had become retitled "Soviet Darwinism".[66] Timiriazev was apparently the only leading Russian Darwinist to engage explicitly with the Bolsheviks and with Marxism after the 1917 Revolution. In the outcomes of the increasingly polemical debates of the 1920s–1930s, he had been attributed the status of "founding father" of both Russian and Soviet Darwinism, and was thus a potentially crucial reference point for politically correct public expositions of Soviet Darwinism.[67]

Vatagin's 1948 bust of Timiriazev (Fig. 1), although a dramatic and heroic image, seems the least convincing of the three busts, in relation to the contemporary demands of Socialist Realism. There is a level of apparent accuracy with regard to the hairstyle and physiognomy in comparison with available photographs of Timiriazev. But there also seems to be an element of caricature regarding the simplified and angular delineation of the facial features and smoothed-over hair, which does not equate comfortably with the contemporary Socialist Realist requirements for naturalistic representation. This nuance may derive from Vatagin's imagination and memories of his acquaintance with Timiriazev as a lecturer during his own studies as a student of biology at Moscow University.[68] It, nevertheless, seems to imbue the image symbolically with an utopian spirit of "militant materialism" regarding the relationship between Marxism and Darwinism, that was popularly associated with Timiriaziev.[69] Kots' ascription of "militant Darwinism" to Timiriazev in his biography of Vatagin offers yet another example of Kots' assiduously self-conscious references to the developing lexicon of Soviet Darwinism.[70]

Vatagin's sculptural images of Lysenko and Michurin shown in Fig. 2 appear to be more convincing in relation to the contemporary rules about Soviet fine art practices. First, because they seem to have greater resemblance to currently available photographic representations. Second, because they also replicate some of the visual tropes associated with the acclaim of these individuals in the Soviet media. The bust of Lysenko, for example, can be seen to emphasise the taut facial muscles and hawk-like

Fig. 2 Photograph of Dmitri Fedulov with display of variation in stuffed domestic and farmed fur-bearing animals, overlooked by busts of Lysenko and Michurin with a quote from Michurin, c.1948, in Kiril Nasedkin, "Taxidermy in Russia by example of the Darwin Museum", International Committee for Museums and collections of natural history, ICOM-NATHIST, 2006, newsletter no.21, January 2007. The online version of this article is no longer available on the ICOM website, but a copy of it has been saved on the "Taxidermy for Cash" website: http://www.taxidermy4cash.com/moscow.html, accessed 16/03/13

gaze evident in media photographs. But the bust also shows him to be more craggily handsome than the photographs, relating to the romantic and idealising element of contemporary Socialist Realist art practice.[71] Likewise, the sculptural bust of Michurin derived from and replicated Soviet media images that emphasised the physical effects of Michurin's work outdoors in nature—the deeply wrinkled skin on his face and forehead—as well as the trademark hat and crumpled suit.

Another part of Kots' plan was to refocus the displays of variety and variation in wild and farmed fur-bearing animals, to demonstrate Michurinist concepts regarding the effect of environment and hybridisation.[72] The photograph reproduced in Fig. 2 is part of the museum's painstaking visual record of displays and activities, used to demonstrate the institution's scientific and political correctness, as well as its significance to its funders. It shows Dmitri Fedulov,[73] the museum's specialist in taxidermy of small, wild and domestic animals, with a display of some of his sculptural creations. According to the current Darwin Museum's Director of Development, Kiril Nasledkin, the foxes' skins were from Soviet fur farms.[74] In relation to this, the image partially can be seen to bear witness to the contribution of the Darwin Museum's professional research to the Soviet economy since the early 1920s, through its close relationship with the fur trade. This was something that Kots was careful to stress in his speech to the staff in 1948 as an evidence of the practical contribution of the Darwin Museum to the state.[75] The stress on the practical uses of Darwinian science was, in itself, an implicit alignment with Lysenko's emphasis on the need for the practical application of Soviet Darwinism.[76]

The carefully posed image can also be argued to exemplify how the shift from a representation of "reactionary" genetics to one of Creative Darwinism could be accomplished relatively painlessly by the museum because of its "laconic" display policy. The photograph shows a display of stuffed foxes and rabbits relating to the Darwin Museum's research interest in variety and variation. This had been restaged using Vatagin's new busts of Lysenko and Michurin, and a short but widely circulated quotation from Michurin that Lysenko's speech had emphatically identified as Michurin's motto.[77] The quotation reads: "We cannot wait for favors from nature. It is our task to wrest them from her".[78] This statement referred explicitly to the Michurinist idea that organisms placed in new environments could be forced to transform, and also, implicitly to Lysenko's belief that artificial, as much as natural hybridisation, was to be seen as an element of natural selection. Fedulov's pose appears to offer suitable—if somewhat theatrically overdone—homage to Lysenko. The shadow of the bust looms large over the display, as an allusion to, and confirmation of Lysenko's power over Soviet biological science at the time.

Kots' plan for the revised museum display contained two other important ingredients. One of these was the third element of the proposed renovated public expositions—a brand new exhibit, entitled "30 Years of Soviet Cattle-breeding".[79] This would involve the production of coloured tables, genealogies, and pictures, including coverage of the new, high-yielding Kostromskoi breed of horned dairy cattle that had been developed at the Karavaevo Sovkhoz in the Kostroma area. The breed was implicitly referred to in Lysenko's speech,[80] and had featured explicitly in the August 1948 VASKhNIL session discussions as a successful example of the application of Michurinist technique, promoted by the Director of the Kostroma cattle-breeding institute, V.A. Shaumyan.

On the fifth day of the VASKhNIL session, Shaumyan pointedly accused "Mendelist-Morganists" of having "common cause" with the "international reactionary force of bourgeois apologists not only for the immutability of genes but also for the immutability of capitalism". Then, having extolled the virtues of the Kostromskoi breed as having been improved by means of the quality and quantity of feed (not of breeding)—as set out by Lysenko's speech—Shaumyan ended with the assertion: "During our many years of work we have steadily been guided by the teaching of that great transformer of nature, I.V. Michurin and his foremost continuator, T.D. Lysenko".[81]

Clearly, the inclusion of the Kostromskoi cattle in the new display would imply the museum's distance from Mendelism-Morganism and assert its engagement with a topical example of the practical "successes" of Michurinist biology. Moreover, such inclusion could be done while retaining a level of scientific integrity at the museum. Despite contemporary Michurinist explanations of the breed's productivity, this had patently been engineered over 30 or more years by traditional methods of cross-breeding the local cattle with larger, foreign breeds known for their hardiness and high milk yields.[82]

The most ingenious and instantaneous part of Kots' plan for "Michurinizing" the museum display, however, was indicated in a handwritten postscript to the typed draft of his speech to the museum staff. He had, in fact, already commissioned a young museum artist, Viktor Evstaf'ev,[83] *after* Lysenko's initial speech but *before* the end of the VASKhNIL session, to go to Michurinsk from August 5–20, 1948.[84] Evstaf'ev's task was to make "sketches and studies", in preparation for a "special exhibition of pictures—portraits" celebrating the life and work of Michurin.[85]

It is worth noting here that Kots had not thought it necessary to do this in June 1935, when Michurin actually died. At this point, although Lysenko's publications were already referring to Michurin as one of the "greats" of Soviet Darwinism, Michurin was still largely regarded by Soviet biologists as one of the "cranks" of Soviet plant biology.[86]

By 1948, however, Michurin was firmly established as an incontrovertible lynchpin in Lysenko's version of "Soviet Creative Darwinism". Michurin's image now needed to be very prominent in the visual representations of Darwinism offered by the Darwin Museum, in order to retain political credibility and the potential for further state funding. This was particularly important in relation to the petition for a new building sent by Kots to Stalin in 1946 stressing the museum's commitment to

"Soviet Creative Darwinism".[87] The petition had apparently resulted in a resolution of the Council of Ministers RSFSR, which had spoken of the "necessity" of a "speedy decision on the question of constructing a new building for the Darwin Museum".[88] Despite the war damage that made the Darwin Museum less inviting to visitors,[89] there were clearly more pressing issues of reconstruction in Moscow at this point, and the decision about the new building was delayed until 1953.[90]

Viktor Evstaf'ev stayed in Michurinsk beyond his initial remit, returning towards the end of September 1948 with a quantity of small, impressionistic oil sketches. One example of these is a painting of *The House Where Michurin Lived* (Fig. 3), representing the Michurin House Museum in the town of Michurinsk in the Tambov province (formerly Kozlov).[91] By 1949, the studies were being translated into a series of larger paintings all of which were 98 × 75cm in size—that is to say, paintings that were quite large, but not of the larger scale or high level of illusionistic naturalism associated at the time by the Academy of Arts USSR with significant monumental

Fig. 3 V.M. Evstaf'ev, *The House Where Michurin Lived*, 1948, sketch, oil on board, 33 × 46cm. By permission of the State Darwin Museum, Moscow. Photo © author, 2010

Socialist Realist paintings.[92] Rather, these works were potentially positioned as mere illustrations of Michurin's life and not examples of Soviet fine art.[93]

Some of the themes in Evstaf'ev's works paralleled elements of a series of paintings and pastel drawings on the life of Darwin commissioned by Kots from the artist Mikhail Ezuchevskii[94] in the 1920s. For example, Evstaf'ev's *Michurin in the Hot-house*, 1949 (Fig. 4) has echoes of Ezuchevskii's 1920s *Darwin in the Hot-house* (Fig. 5), as if demonstrating how Michurin had been a true disciple of Darwin.

There are, however, important differences between the potential significations of the images. In Evstaf'ev's image, the pair of secateurs depicted in Michurin's left hand refer to his activities as a "hands-on" fruit tree grafter. By contrast, Ezuchevskii's image of Darwin, by its representation of a notebook on the greenhouse bench and labels on plant pots, suggest Darwin's well-known, meticulous, scientific recording of his observations.

Fig. 4 V.M. Evstaf'ev, *Michurin in the Hot-house*, 1949, oil on canvas, 98 × 75cm. By permission of the State Darwin Museum, Moscow. Photo © author, 2010

Fig. 5 M.D. Ezuchevskii, *Darwin in the Hot-house*, 1921–1926, pastel on Whatman paper, 100 × 80cm. By permission of the State Darwin Museum, Moscow. Photo © author, 2010

Evstaf'ev's image of Michurin significantly depicts no paperwork at all. What seems to be implied is that Michurin was essentially a practitioner, not a theorist, working with tacit, rather than theoretical knowledge, in order to produce the practical results within agricultural production for which he was eventually honoured by the state. The nuances of the image aligned closely to the way that Michurin was culturally presented at the time. Lysenko's speech, for instance, had stressed Michurin's practical successes in order to trivialise the theoretical nature of Soviet geneticists' experiments with *Drosophila* (fruit flies).[95] On a more populist level, some works in this series were about Michurin's personal quirks—for example, his predilection for mending clocks and watches. The paintings drew explicit attention to his assumed technological prowess and understanding and implied his properly materialistic, practical attitude to the world.

There were also, inevitably, images such as *Michurin and Kalinin*, 1949 (Fig. 6). Mikhail Kalinin, nominal Soviet head of state 1919–1946 and the only "old Bolshevik" not to be "purged" by Stalin, had some passing

Fig. 6 V.M Evstaf'ev, *Michurin and Kalinin*, 1949, oil on canvas, 98 × 75cm. By permission of the State Darwin Museum, Moscow. Photo © author, 2010

connections with agriculture. The image arguably underlined Lysenko's emphasis on Michurin as having been "discovered" by the leaders of the Soviet Communist Party, and thus, his status as having been approved by both Party and state.[96] Historically, both the extent of Lenin's awareness of Michurin, and the date when Kalinin may have visited Michurin's nursery in Tambov province (either 1919 or 1922) seem unclear.[97] But the myth of Lenin's discovery of Michurin, and Kalinin's accolades about Michurin's work became rooted in public consciousness, both in the USSR and abroad through Aleksandr Dovzhenko's film, *Michurin: A Life in Bloom* (1948), released on January 1, 1949.[98]

One of the last in Evstaf'ev's series of narrative reconstructions of Michurin's life was *Ten year-old Michurin in a Meadow*, 1951. At this point, Evstaf'ev's patience with the series was apparently becoming exhausted. Kots was not happy with the results and was particularly critical of the figure. Two increasingly angry letters from Kots in 1952 indicate that Evstaf'ev was supposed to have done another version of this painting, but it never materialised.

In 1955, Evstaf'ev produced two final images of Michurin, one of which was a more than life-size, three-quarter length portrait (Fig. 7). By its size and the detailed and more finished treatment of the figure, particularly the face, the painting indicates that it was to be read as a monumental Socialist Realist portrait of a Soviet hero. This is also underlined by the depicted medals and the serious, angled gaze that appears to be both inward, showing Michurin's psychological depth and directed beyond the viewer towards the future. Yet the treatment of the riotous blossom is quite brushy and impressionistic, a characteristic that not only speaks of Evstaf'ev's personal stylistic preferences but also of subtle changes in the official Soviet art world.

After the death of Stalin in 1953, the strictures surrounding Socialist Realist painting began to relax. One key indicator of this was the reopening[99] of the Pushkin Museum in Moscow in 1954 with a display that included paintings by the French Impressionists, August Renoir, Claude Monet, and Edgar Degas. In the following year, there was a large exhibition of *French Art from the Fifteenth to the Twentieth Century* containing more Impressionist works, which opened at the Pushkin Museum in November and moved to the Hermitage Museum in Leningrad in 1956. As Susan E. Reid and Alison Hilton have argued, these events offered

Fig. 7 V.M. Evstaf'ev, *Portrait of Ivan Michurin*, 1955, oil on canvas, 145 × 185cm. By permission of the State Darwin Museum, Moscow. Photo © author, 2010

opportunities for Soviet artists to re-engage with aspects of modern Western art.[100] The exhibitions also suggested that elements of European Modernist and Impressionistic styles might now be acceptable, within reason, in Socialist Realism. This was borne out by Igor Grabar's review of the travelling exhibition of French art in 1956, with the proviso that style must not supplant or obscure the content and ideological message that should be central to all Soviet art works.[101] It would seem that a year before Grabar's article was published, Evstaf'ev and Kots were confident that the right balance of content and style had been achieved in the formal portrait of Michurin, and it was duly installed in the centre of a small iconostasis within the museum, framed by Vatagin's busts of Lysenko and Michurin, ironically, just when Lysenko was apparently on his way out.

Between 1948 and 1955, there were also other indicators of the museum's public compliance with Lysenko's ideas. One was Kots' commission to Evstaf'ev in 1949 for a portrait of Professor Semen Chernenko (1887–1974) of the I.V. Michurin Fruit and Vegetable Institute Michurinsk, who had won a Stalin Prize in 1947 for developing new varieties of apples and pears.[102] In 1955, Kots also commissioned from Evstaf'ev a portrait of the behavioural psychologist Ivan Pavlov and possibly also one of Pavlov's tutor Ivan Sechenov.[103] These two works may have been connected to the adoption and, as Krementsov has argued, the relocation of Pavlov in relation to Michurinist biology. If so, the commissions indicate that Kots was still keeping abreast of the finer points of the politicised contemporary discourse on biological science.[104] This possibility is underlined by the fact that in an article for the *VOKS Bulletin*[105] in 1955, Kots specifically identified Pavlov—along with Timiriazev and Michurin—as among the "prominent Russian scientists who accepted and developed Darwinism.".[106]

In relation to this, and again, with reference to Krementsov, it is interesting that Kots' personal library catalogue records the acquisition of Pavlov's *Selected Works* (1951), while Ladygina-Kots' library holds an article linking Pavlov with Timiriazev from this period.[107] Moreover, Kots' notes for a lecture on "Human Evolution in the light of Darwinism", dated May 29, 1950, suggest that he was aware of the Michurinist take on Pavlov even before he bought the books. After the obligatory reference to Engels and denunciations of "reactionary", "foreign" interpretations of Darwin, he gave particular emphasis to the work done by the I.P. Pavlov laboratory "and its school".[108] It is also conceivable that Kots' discussion of early hominids in this lecture may have referenced the book by palaeontologist and Lysenko supporter L.Sh. Davitashvilii, significantly acquired for Kots' own library in 1948.[109]

Certainly, the notes for Kots' lecture on "The Bases of Darwinism and Michurinist Biology" in 1950 were slavishly compliant with the language and tenets of Lysenko's speech. For example, the notes differentiated between "classical Darwinism" and Soviet "Creative Darwinism," specifically emphasising the Michurinist idea of the "unity of organism and environment", and also "factors of training" in relation to matters of artificial selection. It also, unsurprisingly, rejected Darwin's Malthusian idea of "struggle for existence" within species, which in this context would signify full compliance with the Lysenkoist interpretation of Darwin currently supported by the Soviet Communist Party.[110] Kots' article in the 1955 VOKS Bulletin continued to assert the museum's support for "Creative Darwinism", albeit with less of the usual Marxist Darwinist rhetoric.[111]

Within Kots' lifetime, the "Michurinization" of the Darwin Museum appeared to have had the desired result with regard to the long-awaited new building. A further petition to Stalin on September 1, 1952, was followed by an order of the Presidium of the Supreme Soviet no.3902, August 1, 1953, giving a provisional deadline of 1957 for the building's construction.[112] While the deadline was set back to 1960, the future event was triumphantly stated in Kots' 1955 VOKS article.[113]

Regarding the proposed new building, the Darwin Museum archive holds some undated (probably early 1950s), but very interesting, faint pencil-drawn designs by Kots for the proposed "Hall of Darwin and the Russian Darwinists" in the new building. Effectively, in this plan, there were to be two iconostases made up of sculpture busts and narrative paintings facing each other. One would focus on Charles Darwin and Alfred Wallace, the other would relate to Russian Darwinism and include representations of Michurin, Lysenko, Timiriazev and other supporters of Soviet "Creative Darwinism".[114]

Yet, beneath the seemingly productive, protective shield of visual and public discursive compliance with Michurinist biology between the 1930s and the 1950s, there were tiny suggestions of Kots' contrary scientific inclinations. For instance, in an undated private letter addressed to Aleksei Komarov,[115] a painter who did a lot of work for the Darwin Museum in the 1940s, there is a playful hint that neither he nor Komarov were wholeheartedly in tune with Michurinism. In December 1949, Kots had asked Komarov to make a portrait of Lysenko,[116] but Komarov never produced any works about Lysenko or Michurinist biology for the museum. In a subsequent undated letter to Komarov and his wife, however, Kots ironically began a Christmas greeting to them: "To the all-renowned proselytisers of the Great Michurin..."[117] Given the circumstances, this seems to underline Kots' lack of sympathy with the Lysenkoist bias that was currently being visually constructed in the museum displays. In addition, with Kots' connivance, the Darwin Museum also harboured the research of geneticist Piotr Smolin on hybridity and variation during the period.[118] Both examples attest to the strategic public use of the vocabulary of Soviet Darwinism to veil the real inclinations of the museum directorate towards biological and genetic science in the period up to c.1955.

1955–1964: Strategic Equivocation between Michurinism and Genetics

The year 1955 marked the acceleration of the multi-layered and often ambivalent process of change in the USSR precipitated by the death of Stalin on March 5, 1953. In the West, it has become known as "de-Stalinisation," but it was referred to in Soviet public rhetoric more euphemistically as a campaign to ameliorate the damage to the state caused by Stalin's "cult of personality".[119] It was a process that was to have contradictory impacts on the Darwin Museum's products and aspirations.

In the realm of artistic culture, as already noted, de-Stalinisation had an effect on expanding the construct of monumental Socialist Realist painting to include the possibility of introducing aspects of impressionistic stylistic techniques that had been banned since the late 1940s. The process also included the use of the Bolshoi Ballet as a highly politicised cultural ambassador abroad, something that was to have an unfortunate impact on the Darwin Museum's aspirations.[120] This was all part of what has become known as the "Thaw" in the so-called "Cold War," in relation to the sporadic support by the Soviet government of (very relative) cultural freedoms in the USSR associated with the period of Nikita Khruschev's ascent to power (1953–1964).[121]

Within the USSR, one element of de-Stalinisation that became significant for the Darwin Museum was the Amnesty Decree published on March 27, 1953, allowing for the release and rehabilitation of some prisoners who had been the victims of Stalin's "Terror". Importantly, it also allowed for the posthumous rehabilitation of some prisoners who had perished. Although problematic and controversial, this move eventually opened the way for the reinstatement of Soviet genetics.[122]

There was also a new emphasis on developing Soviet science and technology, in order to compete with the West, which led to the reopening of communications between Soviet and foreign scientists.[123] In February 1955, for instance, the Ministry of Higher Education issued a decree declaring that lecturers should pay attention to "world science and technology," rather than just focusing on "patriotic" assertions, which located Russian scientists naturally as the leaders in all fields.[124] Furthermore, in July a plenum of the Central Committee of the Soviet Communist Party exhorted scientists to study the achievements of Western science, particularly where these were in advance of those in the USSR. According to the recent research of historian Michael Froggatt, both of these emphases effectively resulted in the diminution of curricular space dedicated to "Creative Darwinism" in Soviet schools.[125] The decrees may also have influenced the low level of Marxist Darwinist vocabulary in Kots' article in the 1955 VOKS Bulletin.

The year 1956 was a high point in the overall process. It was characterised both by the circulation of Khruschev's "secret speech" denouncing Stalin's wrongdoing,[126] and by Lysenko's temporary loss of the VASKhNIL Presidency.[127] Lysenko was to regain this position, equally temporarily between 1961 and 1962, and his powers were not officially entirely eradicated until early 1965. Nevertheless, with hindsight, 1956 signified the beginning of the end for the reign of "Michurinist biology" and "Creative Darwinism" in Soviet bioscience.

The year 1956 certainly marked the sudden cessation in production of Michurin images at the Darwin Museum. After the final portraits of Michurin in 1955, Evstaf'ev's next commission from the Darwin Museum between 1956 and 1957 was for a series of portraits of "eminent Russian Darwinists" that did not fit with the previous model of Lysenkoist Soviet Darwinism.

Significantly, the commission included images of Nikolai Kolt'stov and Nikolai Vavilov. These were Russian scientists who had fallen victim to Lysenko's climb to power. The paintings, which are still on display at the Darwin Museum, do not look any different from any other highly finished, naturalistic Socialist Realist portraits of the period. They safely accorded with the rules for Socialist Realist art laid down by the new Soviet Academy of Arts during the late 1940s, perhaps particularly because the subject matter was potentially contentious.[128] While the style and technique gave the works an aura of authority, the paintings were not over-assertively large, only 55 × 75cm. In this sense, they were not definitively monumental in relation to the continuing requirements of Soviet Socialist Realist art practice, and thus not affording the subjects the highest heroic status. The sheer fact of the commissioning and production of these works, however, is remarkable, given the recent context of the museum's public display of close attention to the fine detail of Lysenko's 1948 speech.

Regarding the portrait of Nikolai Kol'tsov executed by Evstaf'ev in 1956 (Fig. 8), both Kol'tsov and Lysenko had been nominated for full membership of the Academy of Sciences USSR in 1939. Kol'tsov, however, was defeated in the elections due to a vicious public smear campaign by Lysenko's supporters, and died possibly in mysterious circumstances in 1940, followed the next day by the "suicide" of his wife.[129] That he was, nevertheless, still castigated in Lysenko's 1948 speech suggests that at this point, Lysenko still regarded Kol'tsov's writings on genetics as having a potentially dangerous influence on the Soviet scientific community.[130]

The date of the portrait seems to refer to the beginning of an ultimately successful move to reinstate Kol'tsov into the Soviet scientific cannon in which Kots was tentatively involved. In 1958, for instance, Kots wrote to Professor N.P. Dubinin, another target for criticism in Lysenko's speech,[131] who eventually took over the directorship of a revamped Institute of Genetics from Lysenko in 1965, asking for pre-Revolutionary photographs of Kol'tsov.[132] Kots' request was relatively benign. After all, Kol'tsov had recommended Kots for the lecturing post at the Women's Higher Courses Institute in 1907 and had been a fellow lecturer there in the pre-revolutionary period. Kots' stated intention was to write a "necrology" of eminent Russian scientists of this era.[133]

By implication, this would have included those who had been excluded from the history of Soviet science by the official adoption of Michurinism. However, requesting pre-Revolutionary images of Kol'tsov would have suggested that Kots' necrology would gloss over not only Kol'tsov's contributions to Soviet genetics, but also the politically embarrassing facts of Kol'tsov's leading role in the Soviet eugenics movement in the 1920s, and the waves of denunciations related to this, both in the late 1930s and in Lysenko's 1948 speech. From the evidence in the Darwin Museum archives, it is clear that Dubinin supplied the images, but Kots' intention was unfulfilled.[134] Perhaps it was too soon for a celebratory publication that would so seriously challenge the canon of Soviet bioscience as revised by Lysenko. Yet also, perhaps, the request was a covert means to communicate to Dubinin the support of Kots and the Darwin Museum for genetics.

Fig. 8 V.M. Evstaf'ev, *Portrait of Nikolai Kol'tsov*, 1956, oil on canvas, 55 × 75cm. By permission of the State Darwin Museum, Moscow. Photo © author, 2010

The second of the 1956 portraits represented the ornithologist, Professor Mikhail Menzbir, who had been one of the four leading pre-Revolutionary Russian Darwinists given particular prominence after the Revolution (the other three being the brothers A. and W. Kovalevskii and Kliment Timiriazev). Menzbir had been both a teacher, and later, a colleague of Kots, both at the Women's Higher Courses Institute and within the structure of Moscow University after the Revolution.[135] Unlike Timiriaezev, however, Menzbir had not engaged explicitly with Marxist Darwinism. There was, for example, no sign of it in the first Soviet edition of Darwin's collected works published between 1925 and 1929 under his general editorship.[136] While remaining respected after his death in 1935,[137] Menzbir's contribution to Russian Darwinism arguably became somewhat eclipsed in public discourse by the increasing emphasis on Timiriazev from the 1930s onward as the "founding father" of Soviet Darwinism.[138] This trend had culminated in Lysenko's exclusive references to Timiriazev as the sole "true" Russian/Soviet interpreter of Darwin.[139] The production and display of Menzbir's portrait at the Darwin Museum in 1956 could thus be seen as implicitly challenging such a monolithic view and also as an attempt to restore some of Menzbir's deserved status (as well as expressing Kots' possible personal preference for Menzbir's account of Darwinism) into the public arena of discourse on Darwinism at the museum.

The third, somewhat unexpected portrait, was of geneticist Nikolai Vavilov, painted by Evstaf'ev in 1957 (Fig. 9). In 1913–1914 Vavilov had studied with Lysenko's English bugbear, William Bateson.[140] Vavilov was also the most tragic victim of Lysenkoism, having actually supported and indeed, commended Lysenko in the 1930s. By doing so, he contributed unwittingly both to Lysenko's rise in prominence, and to his own descent into hell.[141] Vavilov had also supported Kol'tsov's candidacy for election to the Academy of Sciences USSR in 1939 and, as a result, was subjected to the same smear campaign by Lysenko supporters, leading ultimately to his arrest and sentence as a traitor in 1940. The outcome was that Vavilov was incarcerated in Saratov prison where he died of starvation in January 1943.[142] Vavilov's "rehabilitation" officially began on August 20, 1955, when the Military Collegium of the Soviet Supreme Court ruled that there was no evidence to support his initial conviction.[143] On April 2, 1956, VASKhNIL commissioned the republication of Vavilov's scientific works.[144]

→

Fig. 9 V.M. Evstaf'ev, *Portrait of Nikolai Vavilov*, 1957, oil on canvas, 55 × 75cm. By permission of the State Darwin Museum, Moscow. Photo © author, 2010

The commissioning and display of Vavilov's portrait at the Darwin Museum in 1957, as in the case of the 1956 images of Kol'tsov and Menzbir, suggests Kots' awareness of a contemporary impetus to reinstate these scientists in relation to both the process of de-Stalinisation and Lysenko's apparent loss of power in 1956. It also intimates that the museum's desire to engage with the process was a means to return, legitimately, to its enduring but recently covert engagement with international scientific discourses on evolution, heredity, genetics and variation. Such a return was enabled in a limited way by the opening up of scientific communications between the Darwin Museum and British scientists and institutions in the late 1950s and early 1960s, with the support of VOKS. It was an enterprise conveniently legitimated by a series of significant, Darwin-related anniversaries in 1958, 1959 and 1962.[145]

Nothing that was sent by the Darwin Museum to Britain in this period used the language of Marxist Darwinism, or made any explicit reference to Lysenko and Michurinist biology. This reinforced Kots' position in British eyes as "a real Darwinian"—an accolade articulated in 1956 by the British bioscientist Julian Huxley, who had been in sporadic correspondence with Kots since 1928 and had visited the museum in 1931 and 1945.[146] In turn, this reputation led to the temporary and politically strategic creation of a "Russian Room" at the Darwin house museum at Downe in Kent between c.1962 and c.1964, which was full of materials and art works sent by Kots to the UK on behalf of his museum.[147]

Yet, Kots' notes for a Darwin Museum lecture "On the Theory of Darwin in the Light of Contemporary Genetics," dated August 1958, still paid some lip-service to Marxist Darwinist rhetoric.[148] While there was no explicit reference to Engels, Michurin, or Lysenko, there was a passing mention of "Creative Darwinism" and a cursory critical reference to the anti-Lamarckism of "Neo-Darwinists" such as Weismann, implicitly expressing solidarity with Lysenko's 1948 speech.[149] It is also unclear when the Michurinist iconostasis was finally removed from the museum display, which may not have been until Lysenko had effectively been obliterated from Soviet bioscience early in 1965. Thus, at home Kots was continuing to hedge his bets. This turned out to be immediately wise, but ultimately unwise—in relation to Lysenko's temporary reinstatement as President of VASKhNIL 1961–1962, but subsequent removal from power over Soviet bioscience by early 1965.

In the end, despite all of Kots' stratagems, the promised new building for the Darwin Museum never materialised in his lifetime or indeed in the Soviet period. Once the proposal was personally approved by Stalin in 1952, construction on the designated site was repeatedly delayed.[150] While construction of the building was eventually started, after Kots' death in September 1964 the building was eventually reallocated to a college of choreography attached to the Bolshoi ballet,[151] in order to support a more current polemical cultural resource relating to Soviet "Cold War" propaganda to the West.

Conclusion

Because of Kots' complex stratagems, it is difficult to identify the precise nature of his personal vision of Darwinism over the period under scrutiny. Some of the artworks commissioned and aspects of both his and Ladygina-Kots' libraries indicate a voracious interest in historical and recent Western developments in evolutionary theory, as well as in Russian and Western scientific ideas on heredity and eugenics. At some point after 1920, Kots became a supporter of genetics, possibly retaining, as Darwin had done, a concern for the potential impact of habitat on variety and variation within species.

While these interests were largely, mutely embodied in the Darwin Museum displays, it has been seen that they played an increasingly diminished role in the official presentation of the museum artifacts to the public from the late 1920s to 1955, reappearing partially and very cautiously in the late 1950s during Khrushchev's "Thaw". Instead, the version of Darwinism offered through Kots' lectures and courses in relation to the images seems to have gradually mutated, taking on elements of the developing lexicon of "Marxist Darwinism" particularly from the mid-1930s onwards. Strategically, this was a long-term ploy by Kots to protect the museum, the livelihoods of the staff, himself, and his wife, and also to gain promises of a new museum building from the Soviet state. It was, in effect, a sociopolitical "struggle for existence". This endeavour was implicitly supported by key museum artists, Vasilii Vatagin, Dimitri Fedulov and Viktor Evstaf'ev, without whose efforts, the rapid and effective "Michurinisation" of the museum displays in 1948 and into the early 1950s could not have happened.

In pursuit of the goal of a new museum premises, the articulation of Kots' adoption and adaptation of "Marxist Darwinism" sometimes necessarily conflicted with the actualities of the museum's otherwise cordial links with Western scientists, such as Henry Fairfield Osborn, William Gregory, and

Robert Yerkes in the 1920s–1930s, and Julian Huxley in the 1930s–1950s. It can also be seen in Kots' strategic equivocation in the 1950s, between the un-Michurinised representations of the museum to the British, and the toned-down remnants of "Marxist Darwinist" vocabulary in the 1955 *VOKS Bulletin* and his lectures for Soviet audiences. While the maintenance of the foreign relationships was personally and intellectually important to the Kotses, it was also crucial for the museum to be perceived within the USSR as holding a politically correct position in the contextually changeable Soviet debates on Darwinism.

The latter was particularly imperative in relation to the focal goal of obtaining a new and larger building to house the collection. In relation to this, it is bitterly ironic that this building—promised in 1926, 1946 and 1953—was ultimately given away to the Bolshoi Ballet in the early 1960s. There are three possible reasons for this. First, in the context of de-Stalinisation, the latest promise had been tainted by the hand of Stalin. Second, the proposal had been framed in pro-Lysenko terms, whereas Lysenko had now lost his power over Soviet bioscience. Third, and perhaps most crucially, the productions of the Bolshoi Ballet, and not theorisations of Darwinism, had become a new focus for Soviet Cold War cultural propaganda.

Apart from the commissioning and display of two busts of Engels in 1936, Kots' strategic verbal engagement with "Marxist Darwinism" did not seem to affect the constitution of the museum displays until 1948. This was probably because of Kots' policy of "laconicism" with regard to textual content, and the subsequent reliance of the spectators on the contingent verbal explication of the visual data. The triumph of Lysenko in August 1948, however, may be seen to have had a profound physical impact on the display policies of the Darwin Museum.

This was successfully, strategically managed through Kots' fast footwork in commissioning Vatagin's busts of Lysenko, Michurin, and Timiriazev; the restaging of the display of stuffed fur-bearing animals; the topical new installation about the Kostromskoi breed of cattle; and most particularly, the products of Evstafe'v's work at Michurinsk in 1948. These initial feats were then followed by less sketchy narrative works about Michurin and also by portraits of the Lysenko supporter Chernenko, and of Ivan Pavlov—a relatively new, posthumous recruit to the Lysenkoist cannon.

Kots clearly was no Michurinist but was contextually forced into a high level of compliance with the new, exclusive emphasis on "Michurinist biology" that followed on from the VASKhNIL session of July–August

1948. There is no doubt, however, that by the adoption of the shifting vocabulary of "Marxist Darwinism" and then by the "Michurinising" of the Darwin Museum displays from 1948 to c.1955, Kots and the Darwin Museum unavoidably contributed to the propagandisation and entrenchment within the USSR of Lysenko's "Michurinist Biology". Museum directors, like Kots, did not have much choice in the circumstances, when affiliation to "Marxist Darwinism", and ultimately to Lysenko's definition of Darwinism became compulsory, and noncompliance potentially meant the loss of jobs or funding, or both.

On one level, this case study has suggested that the Darwin Museum's use of artworks and carefully orchestrated displays, discursively framed to suit the shifting cultural, economic, and political contexts in the USSR, can be seen to offer an unusually visual and sensitive charting of the rise and fall of Lysenko. More importantly, the study has offered a vividly ironic example of the operation of the public cultural resource—"Marxist Darwinism"—as a defensive and protective mechanism, and a common negotiating language through which the Darwin Museum was able to access resources from the state to support its educational function in popularising Darwinism.

In the period 1917–1964, the connections between this strategic function and the museum's scientific research can be seen to have become increasingly tenuous. While the latter can be seen to offer an arena of covert resistance to "Marxist Darwinism," effectively, the resources supplied by the state were used to deny rather than promote public access to scholarly scientific research. In relation to this, the case of the Darwin Museum implicitly demonstrates the attendant dangers of transforming complex scientific discourse into an oversimplified set of polemical assertions and formulae. In particular, the story of the Darwin Museum within the period, highlights the loss of science's autonomy and "its embeddedness in a particular mode of production, its hypothetical, relative and provisional character".[152]

Notes

1. Aleksandr Fedorovich Kots (1880–1964): P.V. Bogdanov, ed. *Kollektsii Gosudarstvenogo Darvinovskogo Muzeia*, Moscow: State Darwin Museum, 2002, p. 5.
2. Paul Kammerer, "Das Darwinmuseum zu Moskau", *Monistische Monatshefte*, no.11, 1926, p. 377.

3. Nadezhda Nikolaevna Ladygina-Kots (1889–1963) studied at the Womens' Higher Courses gaining her degree in 1917. She had married Kots in 1911, and set up a zoo-psychological laboratory in 1913. In 1943, she was awarded a PhD in Biological Science on the basis of her previous publications. From 1945 to 1962, she not only worked at the Darwin Museum but also at the psychological section of the Institute of Philosophy attached to the Academy of Sciences USSR. I.P Kalacheva, "Biograficheskaia spravka o fondoobrazatel'e", in I.P. Kalacheva, *Archivnyi fond Ladygina-Kots, Nadezhda Nikolaevna (1889–1963). Opis 1*, Gosudarstvennyi Darvinovskii Muzei, Moskva, October 7, 2010, pp. 3–7
4. A.F.Kots, *Muzei Evoliutsionnoi Istorii Moskovskikh Vysshikh Kursov za 1913–1914 God*, Moscow, 1914 (printed pamphlet, no publisher named). Arkhiv Gosudarstvennogo Darvinovskogo Muzeia (AGDM), f.12430, o.32, ed.khr.165, 16 pp.
5. A.F. Kots, "1907–1917", in Anna Kliukina et al., *Darvinovskii Muzei. 100 Let so dnia osnovaniia 1907–2007*, Moscow, Izdatel'skaia programma interrosa, 2007, pp. 18–19.
6. Bogdanov, p.7. Vasilli Aleksandrovich Vatagin (1883–1969), also trained as an animal illustrator under the leading Russian exponent of this craft, N.A. Martynov, and illustrated the work of the prominent ornithologist and Darwinist Mikhail Menzbir, he also studied with the Russian Symbolist artist Kontantin Iuon: V. Udal'tsova, "Vasilli Vatatagin," *Gosudarstvennyi Darvinovskii Muzei Moskva, Sokrovishcha Russkogo Iskusstva*, Moscow, Belyi gorod/ Gosudarstvennyi Darvinovskii Muzei, 2007, pp. 13–15.
7. Bogdanov, p. 6.
8. Filip Evtikhievich Fedulov (1881–1961), studied with and then worked for the internationally known taxidermist F.K. Lorenz in Moscow and met Kots at Lorenz's studio. He was conscripted into the Imperial cavalry and fought in World War I 1914–1918. For the rest of his life he was the chief taxidermist for the Darwin Museum. Vera Udal'tsova, "Taksidermisty Darvinovskogo muzeia", in Kliukina et al., pp. 75–80.
9. Margarethe Vöhringer, "Behavioural Research, the Museum Darwinianum and Evolutionism in Early Soviet Russia", *History and Philosophy of Life Science*, vol.31, 2009, p. 282. NB. Indeed, in 1913, because the Kotses could not afford it, Fedulov had bought for Ladygina-Kots the chimpanzee (Joni) that was to become the basis of her major, internationally known zoo-psychological project.

A.F. Kots, "1907–1917," p. 21; N.N. Ladygina-Kots, *Infant Chimpanzee and Human Child*, (Scientific Memoires of the Museum Darwinianum, Museum Darwinianum, Moscow, 1935, vol.III) translated by B. Vekker, edited by F. De Waal, Oxford, Oxford University Press, 2002.

10. Kammerer, p. 377.
11. State Darwin Museum Report 1996–2000. "From History of the Museum. Chronickle (sic) of Leading Events", p. 1 of 3, http://www.darwin.museum.ru/report/1996-2000-en/en_hist.htm, accessed 16/11/2008.
12. I.P. Kalacheva and A.I. Kliukina, "100 Let zhizni Darvinovskogo myzeia", in N.I. Tregub, ed. *Trudy Gosudarstvennogo Darvinovskogo Muzeia*, Issue XI, *K 100 Letiiu muzeia*, Moscow, Gosudarstvennyi Darvinovskii Muzei, 2007, p. 76.
13. The UK archives include the British Museum of Natural History, London; the Royal College of Surgeons, England, London; English Heritage, Down House, Kent; the Linnaean Society, London; Christ's College Cambridge and the John Innes Centre in Norwich.
14. Nikolai Krementsov, "Darwinism, Marxism and Genetics in the Soviet Union", in Denis Alexander and Ronald Numbers, eds, *Biology and Ideology from Descartes to Dawkins*, Chicago and London, University of Chicago Press, 2010, pp. 215–246.
15. Ibid., p. 216.
16. Ibid.
17. Ibid.
18. Krementsov, "Darwinism Marxism and Genetics in the Soviet Union", pp. 221–222.
19. Friedrich Engels, "The Part Played by Labour in the Transition from Ape to Man", (1876), trans. Clemens Dutt, Moscow, Progress Publishers, 1934.
20. Friedrich Engels, Letter to Friedrich Albert Lange in Duisburg, *Marx and Engels. Selected Correspondence 1844–1895*, in S.W. Ryazanskaya, ed., translated by I. Lasker, Moscow: Progress Publishers (1955) 1975, pp. 161–162; Friedrich Engels, *Dialectics of Nature*, Moscow: Progress Publishers, 1954, p. 41. Karl Marx letter to Engels in Manchester, London, January 7, 1851, *Marx and Engels. Selected Correspondence*, p. 47; Marx, Letter to Engels in Manchester, London, June 18, 1862, ibid., p. 120; Marx letter to Ludwig Kugelmann in Hanover, London, June 27, 1870, ibid.,

p. 225; Karl Marx, *Capital*, translated by E. and C. Paul, 2 vols, London and New York, (1957) 1972, vol.1, 393, vol.2, p. 574.

21. T.D. Lysenko, "The Situation in Biological Science", address given to the VASKhNIL Session, 31 July 1948, in *Agrobiology. Essays on Problems of Genetics, Plant Breeding and Seed Growing*, Moscow, Foreign Languages Publishing House, 1954, pp. 516–517. For Stalin's editorial input to Lysenko's speech see: Kiril O. Rossianov, "Editing Nature. Joseph Stalin and the New Soviet Biology," *Isis*, vol.84, no.4, December 1983, pp. 728–745, particularly p. 731.
22. Alexander Vucinich, *Darwin in Russian Thought*, Berkeley, Los Angeles, Oxford, University of California Press, 1989, p. 384.
23. N.N. Ladygina-Kots, *Prisposobitel'nye motornye navyki makaka v ysloviiakh eksperimenta: k voprosu o "trudovykh protsessakh" nizshikh obezian'*, Trudy zoopsikhologicheskoi laboratorii, gosudarstvennyi darvinovskii muzei, Moscow, 1928, pp. 351–352.
24. S. Sobol', "Bor'ba za sushchestvovanie," in *Bolshaiia Sovetskaiia Entsiklopedia*, Moscow, Aktsionernoe obshchestvo sovetskaiia entsiklopedia, vol.7, 1927, pp. 204–214.
25. I.S. Agol, "Darvin i Darvinizm," VI, "Predely primeneniia darvinizma", *Bolshaia Sovetskaia Entsiklopediia*, Moscow, Aktsionernoe obshchestvo sovetskaia entsiklopediia, 1930, vol.20, pp. 466–470.
26. A.F. Kots, "Proiskhozhdenie cheloveka (10 chasov) metodolicheskii plan", no date, Arkhiv Gosudarstvennogo Darvinovskogo Muzeia (AGDM), f.10141, o.578, ed.khr.96, pp. 1–4.
27. N. Krementsov, *Stalinist Science*, Princeton, Princeton University Press, 1997, p. 59; Kaganov, "Evoliotionnaia teoriia", pp. 54–86.
28. Ibid., pp. 59–61. Kots' personal library contains a number of Lysenko's publications from the mid-late 1930s, but it is unclear when they were given to him by a certain V. Nikolaev. It also contains Prezent's anthology *Krestomatiia po evoliutsionnom ucheniu*, Leningrad, Leningrad University, 1934. A.F. Kots, "Inventarnaia kniga", no 5, AGDM, f.1243, o.21, ed.khr.106.
29. A.F. Kots, "Proiskhozhdenie cheloveka (10 chasov) metodolicheskii plan". The "recommended reading for teachers" on p.4 lists nothing later than the Party publications, *Uchenie Darvina i marksizm leninizm*, Moscow, Gosizdat, 1932, and *Marks, Lenin, Engels o biologii*, Moscow, Partizdat, 1933, suggesting that the course plan was written between 1933 and early 1935 at the latest.
30. David Joravsky, *The Lysenko Affair*, Chicago, University of Chicago Press, 1986, p. 423, fn. 62.

31. V.M. Kaganov, "Evoliotsionnaia teoriia", *Bolshaiia Sovetskaiia Entsiklopedia*, Moscow, Aktsionernoe obshchestvo sovetskaiia entsiklopedia, 1935, vol.63, pp. 54–86.
32. A.F. Kots letter to W.K. Gregory, November 9, 1934, AGDM, f.12497, o.1079, ed.khr.1239; W.K. Gregory Letter to A.E. Kohts, January 7, 1935, William King Gregory Correspondence 1920–1947, Ms. G7441, American Museum of Natural History Archives (AMNHA).
33. Correspondence of A.F. Kots with colleagues at the American Museum of Natural History: f.376, 492–494, o.10141, ed.khr.1239 and f.1077–1081, o.12497, ed.khr.1239.
34. The donation of the Galton bust is acknowledged in, "Part II. The Exhibit", in *A Decade of Progress. Third International Congress of Eugenics, American Museum of Natural History, New York City, August 22-September 23, 1932*, Baltimore, The Williams and Wilkins Company, 1934, p. 487.
35. Correspondence between A.F. Kots, N.N. Ladygina- Kots and R. Yerkes, AGDM f.12497, o. 650, 651, 653–655, 647–649, ed. khr.1304; f.10141, o. 502–504, ed. khr.1304.
36. Listed in Kots, "Inventarnaia kniga", no 5, AGDM, f.1243, o.21, ed.khr.106.
37. V.A. Vatagin, *Vospominaniia. Zapiski animalista. Stat'i*, Moscow, Sovetskii khudozhnik, 1980, pp. 21, 86. Vatagin notes that this was later replaced with a statue of Michurin by Mikhail Manizer. Kots was director of the Zoological gardens 1920–1923: Vohringer, p. 282.
38. P.N. Yakovlev, "Preface", *I.V. Michurin Selected Works*, Foreign Languages Publishing House, Moscow 1949, http://www.marxists.org/subject/science/essays/preface.htm, accessed 20/12/14; T.D. Lysenko, "Why Bourgeois Science is up in Arms Against Soviet Scientists" (*Literaturnaia gazeta*, October 18, 1947), in Lysenko, *Agrobiology*, pp. 511–514.
39. S. Sobol, "Bor'ba za sushchestvovanie", *Bolshaiia Sovetskaiia Entsiklopedia*, Moscow, Aktsionernoe obshchestvo sovetskaiia entsiklopedia, 1927, vol.7, pp. 203–214.
40. A.E. Gaissinovitch, "The Origins of Soviet Genetics and the Struggle with Lamarckism", trans M.B. Adams, *Journal of the History of Biology*, vol.13, no.1, 1980, pp. 1–51.
41. Ibid.
42. Paul Kammerer, Letter to N.N. Ladygina-Kots, June 19, 1926, AGDM f.11. o.1, ed.khr.2383, pp. 1–2.
43. Kammerer, "Das Darwinmuseum zu Moskau," pp. 377–382.

44. A.S. Serebrovskii, Letter to A.F. Kots, June 10, 1920, AGDM, f.10141, o.253, ed.khr.1026.
45. Vatagin, *Vospominaniia*, pp. 10, 26–28.
46. T. Dobzhansky, "The Birth of the Genetic Theory of Evolution in the Soviet Union in the 1920s," in E. Mayr and W.B. Provine, eds, *The Evolutionary Synthesis: Perspectives on the Unification of Biology*, Cambridge, Mass.: Harvard University Press, 1982, pp. 229–342.
47. Kots, "Vasilii Alekseevich Vatagin i ego raboty v Darvinovskom muzee 1902–1952", pp. 38, 40.
48. A.F. Kots, "The Museum Darwinianum on the Fiftieth Anniversary of its Foundation by Professor Alexander Kohts, Dr. Sc (Biology) Founder and Director of the Museum Darwinianum," *VOKS Bulletin*, no.6 (95), November–December 1955, p. 28, in the English Heritage, Down House Archive (EHDHA), P2/37 (88203381, 88203383, 88203390). NB. 1955 was not actually the 50th anniversary of the Museum, but in the 1950s–1960s Kots' dating of its foundation became ever earlier. Kots also defended this "laconical" approach in other unpublished papers: A.F. Kots, "Ob udarnykh metodakh ekspozitsii", paper given at the first museum conference, December 1930, AGDM, f.10141, o.8, ed.khr.379, pp. 1–30 and Kots,"Vasilii Alekseevich Vatagin i ego raboty v Darvinovskom muzee 1902–1952," p. 62.
49. A.F. Kots, "Doklad na rasshirennom zasedanii itogam sessii VASKhNIL po dokladu akademika Lysenko 'o polozhenii v biologicheskoi nauke'", August 12, 1948, AGDM, f.10141, o.101, ed.khr.33, pp. 1–3; A.F. Kots, "Ob itogakh sessii vsesoiuznoi Akademii s.kh. Nauk po dokladu Akademika Lysenko 'O polozhenii biolog. Nauka' v primenenii k eksponature Gosudarstvennogo Darvinovskogo muzeia", August 12, 1948, AGDM, f.1014, o.101, ed.khr.33, pp. 1–6.
50. Ibid. The directive is mentioned in both speeches.
51. Kots, "Ob itogakh sessii vsesoiuznoi Akademii s.kh. Nauk doklad na rasshirennom zasedanii itogam sessii VASKhNIL po dokladu akademik Lysenko 'o polozhenii v biologicheskoi nauke'", p. 1.
52. Ibid.
53. A.F.Kots, 4 notebooks, AGDM, f.1243, o.0.21, ed.khr.107; Inventory of works by foreign authors in N.N. Ladygina-Kots' personal library, listed in I.P. Kalacheva, *Archivnyi fond Ladygina-Kots, Nadezhda Nikolaevna (1889–1963)*. Opis 1, October 7, 2010, pp. 286–288.

54. Bateson was apparently made an honorary member of the Soviet Academy of Sciences in 1924 but never received any documents confirming this: The John Innes Centre Archive, Norwich (JICA), William Bateson Letters, vol.9, 1198, Bateson Letter to N. Vavilov, April 30, 1924; 1201 Bateson Letter to Vavilov, September 15, 1924.
55. T.D. Lysenko, "The Situation in Biological Science", pp. 517–27, 529, 544–545, 548, 560.
56. In a letter from Lysenko to Stalin in October 1947, Lysenko asserted that "Mendelism Morganism, Weismanist neo-Darwinism" is a "bourgeois metaphysical science" for " reactionary eugenics, racism and other purposes": cited in Ethan Pollock, *Stalin and the Soviet Science Wars*, Princeton, New Jersey: Princeton University Press, 2006, p. 47, fn.20.
57. Lysenko, "The Situation in Soviet Biology", p. 524.
58. Kol'tsov was a family friend and Mukhina's husband, Aleksandr Zamkov, worked with him at the Moscow Institute for Experimental Biology in the 1920s. While Zamkov's concerns were with "rejuvenation" techniques, these can be seen to be an element of the softer, "euthenic" side of Russian eugenics discourse: Pat Simpson, "The New Person's body: drapery and disclosure in Stalinist monumental sculpture", *Sculpture Journal*, 2006, vol.15, no 1, pp. 36–49. See also N. Krementsov, "Eugenics in Russia and the Soviet Union", in Alison Bashford and Philippa Levine, *The Oxford Handbook of The History of Eugenics*, Oxford, New York etc.: Oxford University Press, 2010, p. 421.
59. A.F. Kots, "O printsipakh ekspozitsii i kul'tmassovoi raboty gosudarstvennogo darvinovskogo muzeia," no date (1948–c1955?), AGDM, f.10141, o.190, ed.khr.278 (38 pages), pp. 8, 9, 14.
60. However, some of the images previously on display, in particular a portrait of Mendel by M.D. Ezuchevskii, now had to be hidden away: I.P. Kalecheva, and A.I. Kliukina, "100 let zhizni darvinovskogo muzeia," in N.I. Tregub, ed. *Trudy Gosudarstvennogo Darvinovskogo Muzeia*, Issue XI, *K 100 Letiiu muzeia*, Moscow, Gosudarstvennyi Darvinovskii Muzei, 2007, p. 95.
61. A.F. Kots, "Vasilii Alekseevich Vatagin i ego raboty v Darvinovskom muzee 1902–1952", no date, AGDM, f.10141, o.623, ed.khr.215, p. 33.

62. Lysenko, "The Situation in Biological Science," pp. 532ff.
63. Kalecheva and Kliukina, "100 Let zhizni darvinovskogo muzeia", p. 95.
64. Kots, "Doklad na passhirennom zasedanii itogam sessii VASKhNIL po dokladu akademik Lysenko 'o polozhenii v biologicheskoi nauke'," p. 3.
65. Kots, "Vasilii Alekseevich Vatagin i ego raboty v Darvinovskom muzee 1902–1952", p. 43.
66. Krementsov, "Darwinism, Marxism and Genetics in the Soviet Union," p. 239.
67. Ibid., pp. 217–18, 222–223.
68. Vatagin, *Vospominaniia*, p. 17, 24.
69. Krementsov, "Darwinism, Marxism and Genetics in the Soviet Union", p. 222.
70. Kots, "Vasilii Alekseevich Vatagin i ego raboty v darvinovskom muzee 1902–1952", p. 36.
71. A.A. Zhdanov, "Soviet Literature-The Richest in Ideas. The Most Advanced Literature," in H.G. Scott, ed., Soviet Writers' Congress 1934: The Debate on Socialist Realism and Modernism, London, Lawrence and Wishart (1935) 1977, pp. 21–22.
72. Kots, "Ob itogakh sessii vsesoiuznoi Akademii s.kh. Nauk po dokladu Akademika Lysenko 'O polozhenii biolog. Nauka' v primenenii k eksponature gosudarstvennogo darvinovskogo muzeia,'" pp. 3–4.
73. Dmitri Iakovlevich Fedulov (1896–1985), was Filipp Fedulov's nephew and also studied taxidermy with F.K. Lorenz in Moscow. He began working for the Darwin Museum in 1914, but was conscripted and sent to the front in 1915. When he was demobbed, he served in the Red Army from the end of 1918 to the beginning of 1922, then returned to work for the Darwin Museum. Vera Udal'tsova, "Taksidermisty Darvinovskogo muzeia," in Kliukina et al., *Darvinovskii Muzei. 100 let so dnia osnovaniia 1907–2007*, p. 80.
74. Kiril Nasedkin, "Taxidermy in Russia by example of the Darwin Museum", International Committee for Museums and collections of natural history, ICOM-NATHIST, 2006, newsletter no.21, January 2007. The online version of this article is no longer available on the ICOM website, but a copy of it has been saved on the "Taxidermy for Cash" website: http://www.taxidermy4cash.com/moscow.html, accessed 16/03/13.

75. Kots, "Ob itogakh sessii vsesoiuznoi akademii s.kh. nauk po dokladu akademika Lysenko 'O polozhenii biolog. Nauka' v primenenii k eksponature gosudarstvennogo darvinovskogo muzeia,'" pp. 3–4.
76. Lysenko, "The Situation in Biological Science", pp. 515, 520, 522, 542–543.
77. Ibid., pp. 532, 552.
78. My translation (PS).
79. Kots, "Ob itogakh sessii vsesoiuznoi akademii s.kh. nauk po dokladu akademika Lysenko 'O polozhenii biolog. Nauka' v primenenii k eksponature gosudarstvennogo darvinovskogo muzeia," p. 5.
80. Lysenko, "The Situation in Biological Science," pp. 538–540.
81. *The Situation in Biological Science. Proceedings of the Lenin Academy of Agricultural Science USSR. Session 31. August-September 1948. Verbatim Report*, Foreign Languages Publishing House: Moscow, 1949, pp. 252–260. See also Raissa L. Berg, "On the History of Genetics in the Soviet Union. Science and Politics: The Insight of a Witness", Final Report to the National Council for Soviet and East European Research, Washington University, August 1, 1983, pp. 33–34, http://www.ucis.pitt.edu/nceeer/1983-627-5-Berg.pdf , accessed 18/05/13; D. Lecourt, *Proletarian Science*, p. 98.
82. See S.I. Steiman, *Kak sozdano rekordnoe karavaevskoe stado*, Moscow, Ogiz-sel'khozgiz gosudarstvennoe izdatel'stvo Sel'skokhoziastvennoi literatury, 1948.
83. V.M. Evstaf'ev (1916–1990s) was a painter and sculptor who worked for the Darwin Museum between the 1940s and mid-1950s. Little seems to be available about his career and training, and it may be significant that none of the materials published by the Darwin Museum make any but passing reference to him, perhaps because of his role in "Michurinising" the displays. In the Darwin Museum's current display about the museum's history, the only information on him is that he did a series of works on the life of Darwin.
84. Kots, "Ob itogakh sessii vsesoiuznoi akademii s.kh. nauk po dokladu akademika Lysenko 'O polozhenii biolog. Nauka' v primenenii k eksponature gosudarstvennogo darvinovskogo muzeia", p. 6.
85. Ibid., p. 6. Given the Soviet/Russian obsession with anniversary celebrations, it is likely that this exhibition coincided with the anniversary of Michurin's birth on October 27, 1855.
86. Joravsky, *The Lysenko Affair*, pp. 40–53.

87. A.F. Kots, draft letter to J.V. Stalin, 1946, AGDM, f.12497, o.378, ed. khr.1039, pp. 1–3.
88. Kalecheva, and Kliukina, "100 Let zhizni Darvinovskogo myzeia", p. 95.
89. Ibid., pp. 88–89.
90. Ibid., pp. 102–103.
91. The change of name was made in 1932: N. Bakharev, "Michurin, Ivan Vladimirovich", *Bolshaia Sovetskaia Entsiklopediia*, Moscow, Aktsionernoe obshchestvo sovetskaia entsiklopediia, 1938, vol.37, p 536.
92. For detailed information about the strict rules of Socialist Realism in this period see for eample: B. Ioganson, "O merakh ulucheniia uchebnykh-metodolicheskoi raboti v uchebnykh zavedeniiakh akademii khudozhestva", in P. Sysoev and A Kuznetsov, eds, *Akademiia khudozhestv SSSR, pervaia i vtoraia sessii*, Akademiia khudozhestv SSSR, Moscow, 1949, Session 2, May 20–27 1948, p. 103; A. Gerasimov, "Sovetskoe izobrazitel'noe iskusstvo i zadachi akademii khudozhestv SSSR", in ibid., Session 1, November 22–24 1947, pp. 247, 253, 264–265; A. Gerasimov, "O zadachakh sovetskoi khudozhestvennoi kritiki" (Academic Session 3, January-February 1949), in A. Gerasimov, *Za sotsialisticheskii realism*, Moscow, Akademii khudozhestv SSSR, 1952, pp. 138–144; A. Gerasimov, "Vyshe znamia sotsialisticheskogo realizma" (*Izvestiia deputatov trudiashchikhisia SSSR*,no.39, 1949), in ibid., pp. 155, 158; A. Gerasimov, "Sovetskoe izobrazitel'noe iskusstvo i zadachi akademii khudozhestv SSSR", p. 238; Pat Simpson, "On the Margins of Discourse? Visions of New Socialist Woman in Soviet Art 1949–1950," *Art History*, vol.21, no.2, June 1998, pp. 247–267.
93. It would seem from the writings of both Kots and Vatagin, that if adverse criticism was forthcoming, any perceived aesthetic deviations or stylistic infelicities in art work produced for the museum, could be excused on the grounds that it was "illustration" for the purpose of accurate communication of scientific ideas, thus implicitly should not be judged by the standards set for Socialist Realist fine art: Kots, "Vasilii Alekseevich Vatagin i ego raboty v darvinovskom muzee 1902–1952", pp. 47, 52, 54; Vatagin, *Vospominaniia*, p. 7.
94. Mikhail Dimitrevich Ezuchevskii (1880–1928) trained in zoological illustration at N.A. Martynov's studio alongside Vatagin, then studied fine art at the Academie des Beaux Arts in Paris. On his

return to Moscow in 1914, he exhibited at Lemer's gallery in the company of famous Russian artists such as A.M. and K.M Vasnetsov, M.A. Vrubel, and I.E. Repin. At the outbreak of World War I he was conscripted into the army. When he returned in 1918, Vatagin persuaded Kots to invite Ezuchevskii to work for the museum, for which he produced paintings and pastel drawings presenting narratives from the history of the development of the science of natural history, including representations of Saint-Hilaire, Buffon, Goethe, Lamarck, as well as a series on the life of Darwin: V. Udal'tsova, "Mikhail Dimitrivich Ezuchevskii", in *Gosudarstvennyi Darvinovskii Muzei Moskva, Sokrovishcha Russkogo Iskusstva*, p. 28.

95. Lysenko, "The Situation in Biological Science", pp. 530–532.
96. Ibid., p. 544.
97. See David Joravsky, "The First Stage of Michurinism", in John Shelton Curtiss, ed. *Essays in Russian and Soviet History: In Honour of Gerald Tanquary Robinson*, New York, Columbia University Press, 1963, p. 126.
98. See, for example, A.W., "Film 'Life in Bloom', film of life of a Russian scientist is new picture at the Stanley Theatre," *The New York Times*, May 9, 1949, http://movies.nytimes.com/movie/review?res=9B0DE3DC103CE23BBC4153DFB3668382659EDE, accessed 29/07/13.
99. It had been closed in 1947 because of the alleged corrupting, "cosmopolitan" influence of the western Impressionist paintings contained in its collection: M. Cullerne Bown, *Socialist Realist Painting*, New Haven and London, Yale University Press, 1998, pp. 192–195, 222, 224–225, 282–284, 280, 288–289.
100. Susan E. Reid, "Toward a New (Socialist) Realism," in Rosalind P. Blakesley and Susan Reid, eds, *Russian Art and the West: A Century of Dialogue in Painting, Architecture and the Decorative Arts*, Dekalb Illinois, Northern Illinois University Press, 2007, pp. 217–221; Alison Hilton, "*Holiday on the Kolkhoz*: Socialist Realism's Dialogue with Impressionism", in ibid. pp. 195–216.
101. Igor Grabar, "Zametki o zhivopisi", *Literaturnaia gazeta*, no .115, September 27, 1956, p. 3.
102. "Chernenko, Semen Fedorovich", *The Great Soviet Encyclopaedia*, 1979, http://encyclopedia2.thefreedictionary.com/Semen+Fedorovich+Chernenko , accessed 25/03/13.

103. The latter portrait is undated in the AGDM catalogue of art works, so the date is speculative.
104. Krementsov, *Stalinist Science*, p. 287.
105. VOKS, the acronym of the Vsesoiuznoe obshchestvo kul'turnoi sviazi s zagranistei (All-Union Society for Cultural Relations with Foreign Countries), was the official channel for contacts with the West, and the *VOKS Bulletin* was a propaganda vehicle aimed at western cultural organisations.
106. Kots, "The Museum Darwinianum on the Fiftieth Anniversary of its Foundation", p. 28.
107. E.V. Polosatova, "Perepiski I.P. Pavlova s K.A. Timiriazevym", *Izvestiia AN SSSR*, no.4, 1949, Arkhiv Gosudarstvennogo Darvinovskii Muzeia, listed in I.P. Kalacheva, *Archivnyi fond Ladygina-Kots, Nadezhda Nikolaevny* (*1889–1963*), p. 306.
108. A.F. Kots, "Tezisnoe soderzhanii dvukh lektsii. Lektsiia I 'Osnovy darvinizma i michurinskaia biologiia': Lektsiia II 'Proiskhozhdente cheloveka v svete darvinizma'," May 29, 1950, AGDM, f.12430, o.6, ed.khr.39, pp. 1–2.
109. L. Sh. Davitashvili, *Istoriia paleontologie ot Darvina do nashikh dnei*, Moscow-Leningrad, Akademii nauk SSSR, 1948, listed in AGDM, f.1243, o.21, ed.khr.107.
110. Kots, "Tezisnoe soderzhanii dvukh lektsii."
111. Kots, "The Museum Darwinianum on the Fiftieth Anniversary of its Foundation," pp. 26–32.
112. Kalacheva and Kliukina, "100 let zhizni darvinovskogo muzeia", pp. 102–106.
113. Kots, "The Museum Darwinianum on the Fiftieth Anniversary of its Foundation," p. 30.
114. A.F. Kots, "Zal Darvin i russkie darvinisty", no date (early 1950s), AGDM, f.10141, o.169, ed.khr.378, pp. 1–2 of 2.
115. Aleksei Nikanoforovich Komarov (1879–1977) studied landscape and animal painting at the Moscow School of Painting Sculpture and Architecture with the well-known Russian artist A.S. Stepanov. He worked for the Darwin Museum 1932–1965. Vera Udal'tsova, "Khudozhniki Darvinovskogo muzeia," in Kliukina et al., *Darvinovskii Muzei. 100 let so dnia osnovaniia 1907–2007*, Moscow, pp. 88–89.
116. A.F. Kots Letter to A.N. Komarov, December 17, 1949, AGDM, f.12497, o.460, ed.khr.961, pp. 1–2

117. A.F. Kots Letter to A.N. and N.A. Komarov, no date, AGDM, f.12497, o.771, ed.khr.961.
118. Kalacheva and Kliukina, "100 Let zhizni darvinovskogo muzeia", pp. 91–92, 96–97.
119. Polly Jones, "Introduction", in Polly Jones, ed. *The Dilemmas of De-Stalinisation: Negotiating Cultural and Social Change in the Khruschev Era*, London and New York, BASEES/Routledge, 2006, p. 2.
120. Kirsten Hamm, "'The Friendship of Peoples.' Soviet Ballet, Nationalities Policy and the Artistic Media," MA Thesis submitted to the Graduate School of the University of Illinois at Urbana-Champaign, 2009, https://www.ideals.ilinois.edu/bitstream/handle/2142/1432/hamm_kristen.pdf?sequence=2, accessed 25/03/13.
121. Jones, "Introduction", pp. 3, 5.
122. Michael Heller and Aleksandr Nekrich, *Utopia in Power: A History of the USSR from 1917 to the Present*, translated by Phyllis B. Carlos, London, Melbourne, Auckland, Johannesburg, Hutchinson, 1982, p. 526; Miriam Dobson, " 'Show the bandit enemies no mercy'!: amnesty, criminality and public response in 1953", in Jones, ed., pp. 21–37.
123. Nikolai Krementsov, "A 'Second Front' in Soviet Genetics: The International Dimension of the Lysenko Controversy, 1944–1947," Journal of the History of Biology, vol.29, 1996, pp. 229–250; Andrew Bone, "Introduction," in A.Bone, ed. *The Collected Papers of Bertrand Russell*, Abingdon and New York, Routledge, 2003, vol.28, *Man's Peril 1954–1955*, (no page numbers).
124. Michael Froggatt, "Renouncing dogma, teaching utopia: science in schools under Khruschev", in Jones, ed. *The Dilemmas of De-Stalinisation*, p. 255.
125. Ibid.
126. Heller and Nekrich, pp. 530–532.
127. Conway Zirkle, "L'Affaire Lysenko", *Journal of Heredity*, vol.47, no.2, 1956, p. 47.
128. Susan E Reid, "Modernising Socialist Realism in the Khrushchev Thaw: The Struggle for a 'Contemporary Style', in Soviet Art", pp. 209–230,
129. Peter Pringle, *The Murder of Nikolai Vavilov*, New York, Simon and Shuster, 2011, p. 339: Valery N. Soyfer, "The Consequences of

Political Dictatorship for Russian Science", *Nature Reviews Genetics*, no.2, September 2001, pp. 723–729.
130. Lysenko, "The Situation in Biological Science", p. 524.
131. Ibid., pp. 524, 526, 530–532.
132. Dubinin's Cytogenetical Laboratory of Cytology, Histology and Embryology had been abolished as "unscientific and useless" by a resolution of the Presidium of the Academy of Sciences USSR, on August 26, 1948: Julian Huxley, "Soviet Genetics: the Real Issue", (from *Nature*, June 1949) in John McCormick and Mairi MacInnes, eds, *Versions of Censorship*, New Jersey, Transaction Publishers, 2006, pp. 97–98.
133. A.F. Kots, Letter to N.P. Dubinin, December 18, 1958, AGDM, f.10141, o.271, ed.khr.937.
134. Pre-Revolutionary photographs of Kol'tsov at the Women's Higher Courses Institute, AGDM, f.2095, khr.109, box no.22.
135. Vatagin, *Vospominaniia*, pp. 17, 24.
136. M.A. Menzbir, *Polnoe Sobrainie Sochinenii Charlza Darvina*, 14 vols, Moscow and Leningrad, Gosudarstvevennoe izdatel'stvo, 1925–1929.
137. Signified partly by the memorial volume: Gregorii Petrovich Dement'ev, ed. *Pamiati akademika Mikhaila Aleksandrovicha Menzbira* (Sbornnik statei), Moscow and Leningrad, Akademiia nauk USSR, 1937.
138. Krementsov, "Darwinism, Marxism and Genetics in the Soviet Union", p. 232.
139. See for example: T.D. Lysenko, "K.A. Timiryazev and The Tasks of our Agrobiology", (address at meeting of the Academy of Sciences USSR, June 3 1943, to celebrate the centenary of Timiriazev's birth), in *Agrobiology*, pp. 374–389; "The Situation in Biological Science", pp. 517–518, 528.
140. Pringle, *The Murder of Nikolai Vavilov*, pp. 36–37.
141. Nils Roll-Hansen "A New perspective on Lysenko," *Annals of Science*, vol.42, 1985, pp. 262–278.
142. Pringle, *The Murder of Nikolai Vavilov*, p. 310.
143. Yevgenia Albats, *KGB: State within a State*, London, I.B. Tauris, 1995, p. 77.
144. John Saville, "The Twentieth Congress and the British Communist Party," *The Socialist Register*, 1979, p. 2.
145. 1958 was the 150th anniversary of the publication of a joint paper by Charles Darwin and Alfred Russell Wallace to the Linnaean

Society, London (July 1, 1858); 1959 was the centennial anniversary of the publication of *Origin of Species*; 1962 was the ninth anniversary of Darwin's death (1882).

146. J.S. Huxley, Letter to H.J. Muller December 12, 1956, in K.R. Dronamraju, *The Life and Work of Julian Huxley*, Singapore, New Jersey and London, World Scientific Publishing Co., 1993, p. 215.
147. This is the subject of a future publication.
148. A.F. Kots,' O teorii Darvina i sovremennoi genetike', August 1958, AGDM, f.10141, o.475, ed.khr.52, pp. 1–36.
149. Lysenko, "The Situation in Biological Science", p. 521.
150. See, for example, Letter from Sir James Porritt to the Soviet Ambassador, December 7, 1960, The Royal College of Surgeons of England Archive (RCSEA), RCS-MUS/14/9.
151. Kalacheva and Kliukina, "100 let zhizni Gosudarstvennogo Darvinovskom Muzeia", pp. 64–205.
152. Krementsov, "Darwinism, Marxism and Genetics in the Soviet Union," p. 246.

Index

Note: Page numbers followed by "n" refers to notes.

W. deJong-Lambert, N. Krementsov (eds.),
The Lysenko Controversy as a Global Phenomenon, Volume 1,
DOI 10.1007/978-3-319-39176-2

R

The manufacturer's authorised representative in the EU is Springer Nature Customer Service Centre GmbH, Europaplatz 3, 69115 Heidelberg, Germany. If you have any concerns regarding our products, please contact ProductSafety@springernature.com

Printed and bound by CPI Group (UK) Ltd, Croydon, CR0 4YY
30/09/2026
02225649-0002